INTRODUCTION TO
RADAR SYSTEMS

By

K.K. Sharma

Assitant Professor
Deptt. of Electronics & Communication Engg.
LINGAYA's University
Faridabad

S.K. KATARIA & SONS®

Publisher of Engineering Books
4885/109, Prakash Mahal, Dr. Subhash Bhargav Lane,
Opposite Delhi Medical Association, Daryaganj, New Delhi-110002
Phone: +91-11-43551243; Mobile: +91-9871775858, 7678425209
e-mail: katariabook@gmail.com, katariabooks@yahoo.com
Website : www.skkatariaandsons.com

Published by :

S.K. Kataria & Sons®

4885/109, Prakash Mahal, Dr. Subhash Bhargav Lane,
Opposite Delhi Medical Association, Daryaganj, New Delhi-110002
Phone: +91-11-43551243, Mobile: +91-9871775858, 7678425209
e-mail: katariabook@gmail.com, katariabooks@yahoo.com

First Edition: 2012, Reprint: 2026

Price : ₹ 295/-

Laser Typesetting by : AISA Enterprises   *Printed at :* Nisha Enterprises

Preface

This text book, **"Introduction to Radar Systems"** is intended to serve as a one semester text for undergraduate course of Electronics and Communication Engineering students. The purpose of this book is to extend and update the knowledge of the reader to the basic concepts of radar systems. In this book, the fundamental concepts of radar systems are explained in a simple manner.

The book has been written in a very simple language keeping in focus the syllabi prescribed for the subject. All the topics of this book have been illustrated with the photograph and solved problems.

This book is an outgrowth of this need and hope it will be very helpful for the students.

Inspite of sincere efforts there may be some misprints, omission and errors. Any suggestion towards the improvement of this text will be highly appreciated.

Faridabad **K. K. SHARMA**
AUTHOR

Acknowledgements

I take this opportunity to express my gratitude to the respected Late Prof G.V.K. Sinha, founder Chairman of Lingaya's University, Dr. K. K. Aggarwal, Chancellor, Lingaya's University, Dr. P. Gadde, CEO and Pro Chancellor, Lingaya's University and Dr. G.S. Yadava, Pro Vice-Chancellor, Lingaya's University, Faridabad for providing innovative and encouraging environment to work on this project.

I would like to thank to the Management, teachers, students and supporting staff of Lingaya's University. I would like to express my gratitude to all the authors whose work has been freely consulted in the preparation of this text.

I am grateful to my wife Sunita, my daughters Aparna and Vaishali, and son Aman for supporting me to complete this project.

A special thanks to Mr Sanjeev Kataria, S. K. Kataria & Sons, New Delhi and Mr Sachin and Mr Anoop for publishing this book in such a beautiful get-up and well in time.

Lastly, I am grateful to my father and mother without their blessing this book would have not come into existence.

K. K. Sharma

This book is Dedicated to

My Beloved Wife

Sunita

Contents

CHAPTER

1

INTRODUCTION TO RADAR

1.1. INTRODUCTION

INSIDE THIS CHAPTER

Radar is an acronym for **Radio Detection and Ranging**. The term "radio" refers to the use of electromagnetic waves with wavelengths in the so-called radio wave portion of the spectrum, which covers a wide range from 100 km to 1 cm. It is a system used to detect, range (determine the distance) and map objects such as aircraft and rain. Strong radio waves are transmitted, and a receiver listens for reflected echoes. By analysing the reflected signal, the reflector can be located, and sometimes identified. Although the amount of signal returned is tiny, radio signals can easily be detected and amplified.

It can operate in darkness, haze, fog rain and snow. It has ability to measure distance with high accuracy in all weather conditions.

The electronics principle on which radar operates is very similar to the principle of sound-wave reflection. If you shout in the direction of a sound-reflecting object (like a rocky canyon or cave), you will hear an echo. If you know the speed of sound in air, you can then estimate the distance and general direction of the object. The time required for a return echo can be roughly converted to distance if the speed of sound is known.

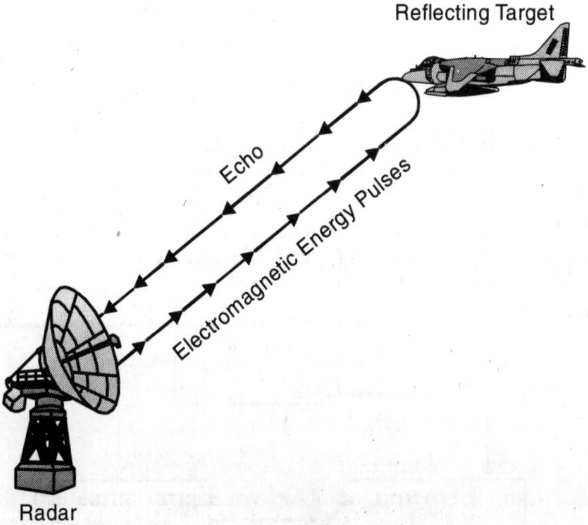

Fig. 1.1. Radar Principle.

Radar uses electromagnetic energy pulses in the same way, as shown in Fig. 1.1. The radio-frequency (rf) energy is transmitted to and reflects from the reflecting object. A small portion of the energy is reflected and returns to the radar set. This returned energy is called an **ECHO**, just as it is in sound terminology. Radar set use the echo to determine the direction and distance of the reflecting object.

Fig. 1.2. An Air Traffic Control Radar.

Figure 1.2 shows the real look of an Air Traffic Control Radar, which is being used as Ground Control Approach Radar at air bases in civil as well as in defense.

1.2. RADAR BLOCK DIAGRAM

A practical radar system requires seven basic components as illustrated below :

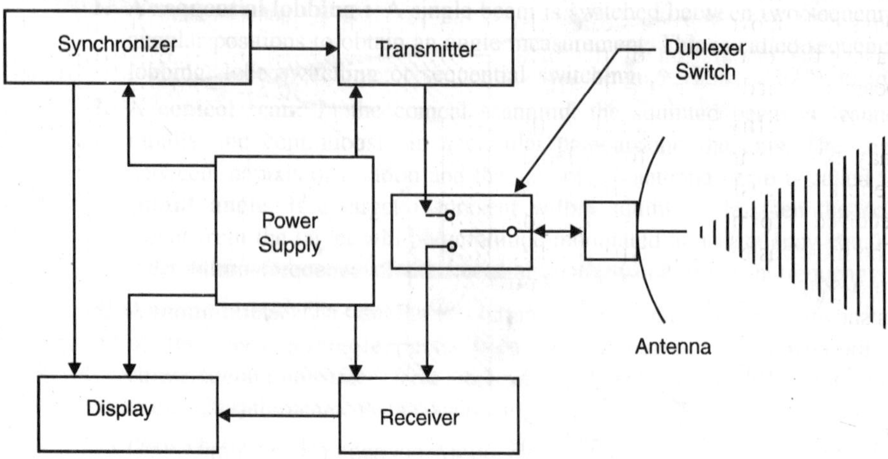

Fig. 1.3. A Practical Simple Radar Block Diagram.

1. Transmitter. The transmitter creates the radio wave to be sent and modulates it to form the pulse train. The transmitter must also amplify the signal to a high power level to provide adequate range. The source of the carrier wave could be a Klystron, Travelling Wave Tube (TWT) or Magnetron. Each has its own characteristics and limitations.

2. Receiver. The receiver is sensitive to the range of frequencies being transmitted and provides amplification of the returned signal. In order to provide the greatest range, the receiver must be very sensitive without introducing excessive noise. The ability to discern a received signal from background noise depends on the signal-to-noise ratio (S/N).

3. Power Supply. The power supply provides the electrical power for all the components. The largest consumer of power is the transmitter, which may require several kW of average power. The actually power transmitted in the pulse may be much greater than 1 kW. The power supply only needs to be able to provide the average amount of power consumed, not the high power level during the actual pulse transmission. Energy can be stored, in a capacitor bank for instance, during the rest time. The stored energy then can be put into the pulse when transmitted, increasing the peak power.

4. Synchronizer. The synchronizer coordinates the timing for range determination. It regulates that rate at which pulses are sent (*i.e.* sets PRF) and resets the timing clock for range determination for each pulse. Signals from the synchronizer are sent simultaneously to the transmitter, which sends a new pulse, and to the display, which resets the return sweep.

5. Duplexer. This is a switch that alternately connects the transmitter or receiver to the antenna. Its purpose is to protect the receiver from the high power output of the transmitter. During the transmission of an outgoing pulse, the duplexer will be aligned to the transmitter for the duration of the pulse, PW. After the pulse has been sent, the duplexer will align the antenna to the receiver. When the next pulse is sent, the duplexer will shift back to the transmitter. A duplexer is not required if the transmitted power is low.

6. Antenna. The antenna takes the radar pulse from the transmitter and puts it into the air. Furthermore, the antenna must focus the energy into a well-defined beam, which increases the power and permits a determination of the direction of the target. The antenna must keep track of its own orientation, which can be accomplished by a synchro-transmitter. There are also antenna systems, which do not physically move but are steered electronically (such as phased array antenna, in these cases, the orientation of the radar beam is already known a priori).

7. Display. The display unit may take a variety of forms but in general is designed to present the received information to an operator. The most basic display type is called an A-scan (amplitude vs. Time delay) as shown in Fig. 1.4. The vertical axis is the strength of the return and the horizontal axis is the time delay, or range. The A-scan provides no information about the direction of the target.

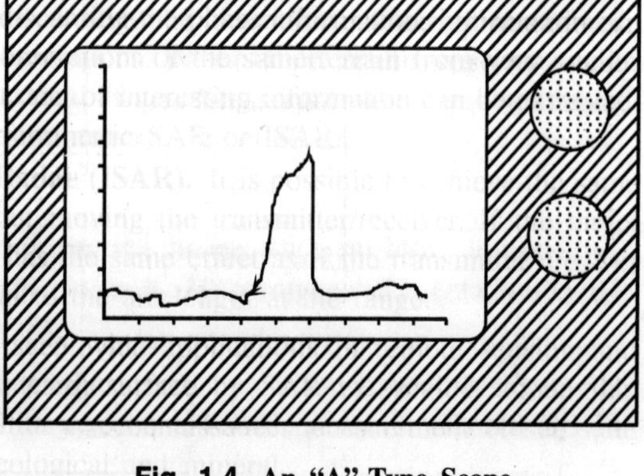

Fig. 1.4. An "A"-Type Scope.

The most common display is the PPI (plan position indicator) as shown in Fig. 1.5. The A-scan information is converted into brightness and then displayed in the same relative direction as the antenna orientation. The result is a top-down view of the situation where range is the distance from the origin. The PPI is perhaps the most natural display for the operator and therefore the most widely used. In both cases, the synchronizer resets the trace for each pulse so that the range information will begin at the origin.

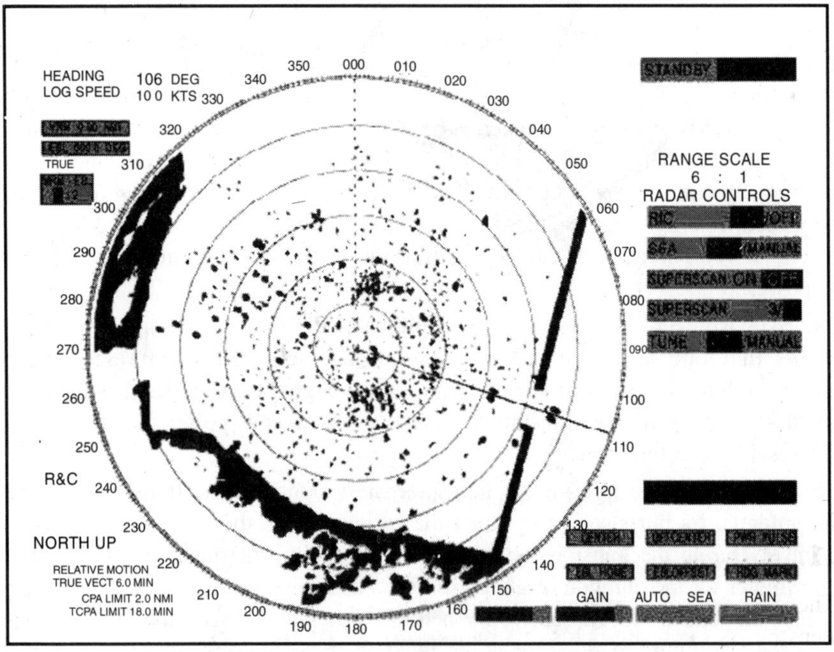

Fig. 1.5. Display In Plan Position Indicator (PPI).

1.3. DETERMINATION OF CO-ORDINATES IN RADAR

Radar requires a more precise reference system. Radar surface angular measurements are normally made in a clockwise direction from TRUE NORTH, as shown in Fig. 1.6, or from the heading line of a ship or aircraft. The surface of the earth is represented by an imaginary flat plane, tangent (or parallel) to the earth's surface at that location. This plane is referred to as the HORIZONTAL PLANE. All angles in the up direction are measured in a second imaginary plane that is perpendicular to the horizontal plane.

This second plane is called the VERTICAL PLANE. The radar location is the center of this coordinate system.

The line from the radar set directly to the object is referred to as the **LINE OF SIGHT (LOS)**. The length of this line is called **RANGE.**

The angle between the horizontal plane and the LOS is the ELEVATION ANGLE. The angle measured clockwise from true north in the horizontal plane is called the **TRUE BEARING or AZIMUTH** angle.

These three coordinates of range, bearing, and elevation describe the location of an object with respect to the antenna.

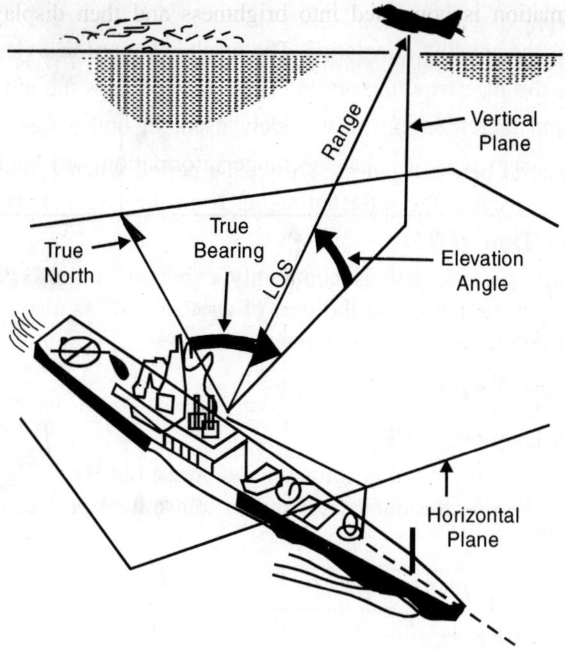

Fig. 1.6. Radar Reference Coordinates.

1.3.1. Range

The detection and ranging part of the radar is accomplished by timing the delay between transmission of a pulse of radio energy and its subsequent return. If the time delay is Δt, then the range may be determined by the simple formula :

$$R = c \times \left(\frac{\Delta t}{2}\right) \qquad ...(1.1)$$

where $c = 3 \times 10^8$ m/s, the speed of light at which all electromagnetic waves propagate. The factor of two in the formula comes from the observation that the radar pulse must travel to the target and back before detection, or twice the range.

A radar pulse train is a type of amplitude modulation of the radar frequency carrier wave, similar to how carrier waves are modulated in communication systems. In this case, the information signal is quite simple, a single pulse repeated at regular intervals. The common radar carrier modulation, known as the **pulse train** is shown below. The common parameters of radar as defined by referring to Fig. 1.7.

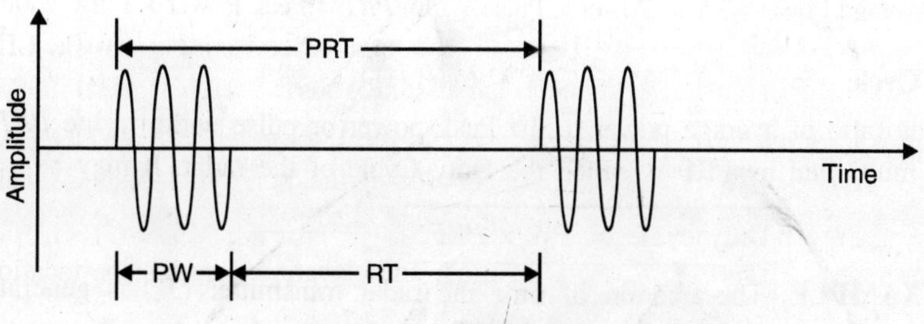

Fig. 1.7. A Radar Pulse.

Common Parameters of Radar Pulse

Pulse Width (PW)

PW has units of time and is commonly expressed in ms. *PW* is the duration of the pulse.

Rest Time (*RT*)

RT is the interval between pulses. This is the period when T_x is silence (not firing) and R_x is ready to receive the reflected signal from the target. It is measured in ms.

Pulse Repetition Time (*PRT*)

PRT has units of time and is commonly expressed in ms. *PRT* is the interval between the start of one pulse and the start of another. *PRT* is also equal to the sum of pulse width and rest time. It can be written as

$$\boxed{PRT = PW + RT}$$

Pulse repetition frequency (*PRF*)

PRF has units of time^{-1} and is commonly expressed in Hz (1 Hz = 1/s) or as pulses per second (pps). *PRF* is the number of pulses transmitted per second and is equal to the inverse of *PRT*.

$$\boxed{PRF = 1/PRP}$$

Radio frequency (RF)

RF has units of time^{-1} or Hz and is commonly expressed in GHz or MHz. *RF* is the frequency of the carrier wave which is being modulated to form the pulse train.

Peak power (P_t)

The power P_t in the radar equation is called by the radar engineer, the Peak Power. The **peak power** as used in the radar equation is not the instantaneous peak power of a sine wave. It is defined as the power averaged over that carrier frequency cycle which occurs at the maximum of the pulse power. Peak power is usually equal to one-half of the maximum instantaneous power.

Average Power (P_{av})

The **average power P_{av}** is defined as the average transmitted power over the pulse repetition time or period. If the transmitted waveform is a train of rectangular pulses of width and pulse repetition period or time.

$$\boxed{PRT = \frac{1}{PRF}} \quad ...(1.2)$$

$$\boxed{P_{av} \text{ (Average Power)} = P_t \times \frac{PW}{PRT} = P_t \times PW \times PRF} \quad ...(1.3)$$

(Average Power = Peak Power × Pulse Width/*PRT* = Peak Power × Pulse width × PRF)

Duty Cycle

The ratio of average power to the Peak power or pulse width to the *PRT* or Pulse width multiplied by PRF is called the Duty Cycle of the radar. It may be written as,

$$\boxed{\text{Duty cycle} = PW \times PRF} \quad ...(1.4)$$

EXAMPLE. The fraction of time the radar transmitter (T_x) is generating pulse power is termed the Duty-Cycle. Thus, in any one second, radar producing 1 μs pulses at a rate of 1000 pulse per sec is "on". The duty cycle may be calculated by using Eqn. (1.4).

$$\text{Duty cycle} = PW \times PRF = 1 \times 10^{-6} \times 1000 = 0.001 \text{ s}$$

The duty cycle is then 0.001, and the average power is given by the product of duty cycle and peak power; thus if the peak power is 500 kW, the average power is then

$$\text{Average Power } (P_{av}) = P_t \times \text{Duty Cycle}$$

$$P_{av} = 500 \times 10^3 \times 0.001$$

$$P_{av} = 500 \text{ W}$$

1.3.2. Bearing

The **TRUE BEARING** (referenced to true north) of a radar target is the angle between true north and a line pointed directly at the target. This angle is measured in the horizontal plane and in a clockwise direction from true north. The bearing angle to the radar target may also be measured in a clockwise direction from the centerline of your own ship or aircraft and is referred to as the **RELATIVE BEARING**. Both true and relative bearing angles are illustrated in Fig. 1.8.

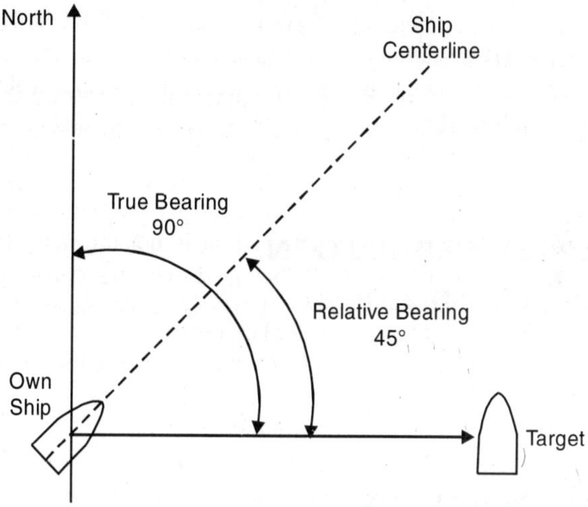

Fig. 1.8. True and Relative Bearings.

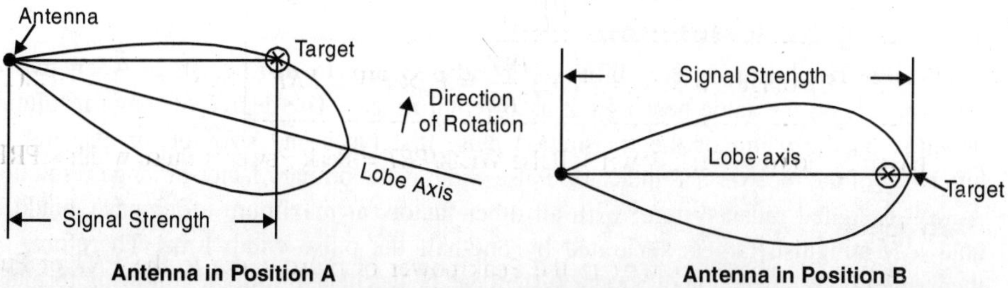

Fig. 1.9. Determination of Bearing.

The antennas of most radar systems are designed to radiate energy in a one-directional lobe or beam that can be moved in bearing simply by moving the antenna. As you can see in Fig. 1.9, the shape of the beam is such that the echo signal strength varies in amplitude as the antenna beam moves across the target. At antenna position A, the echo is minimal; at position B, where the beam axis is pointing directly at the target, the echo strength is maximum. Thus, the bearing angle of the target can be obtained by moving the antenna to the position at which the echo is strongest. In actual practice, search radar antennas move continuously; the point of maximum echo return is determined by the detection circuitry as the beam passes the target or visually by the operator.

Weapons-control and guidance radar systems are positioned to the point of maximum signal return and maintained at that position either manually or by automatic tracking circuits.

1.3.3. Altitude

Many radar systems are designed to determine only the range and bearing of an object. Such radar systems are called **TWO-DIMENSIONAL (2D) radars**. In most cases these systems are further described as **SEARCH RADAR SYSTEMS** and function as early-warning devices that search a fixed volume of space. The range and bearing coordinates provide enough information to place the target in a general area with respect to the radar site and to determine distance, direction of travel, and relative speed.

However, when action must be taken against an airborne target, altitude must be known as well. An altitude is height of the target from the ground plane, sometimes it is considered as the height from the Mean Sea Level (MSL). A search radar system that detects altitude as well as range and bearing is called a **THREE-DIMENSIONAL (3D) radar**. Altitude- or height-finding search radars use a beam that is very narrow in the vertical plane. The beam is scanned in elevation, either mechanically or electronically, to pinpoint targets. Height-finding radar systems that also determine bearing must have a beam that is very narrow in both the vertical and horizontal planes. The display system use for indicating the height of the target is known as **Height Measuring Indicator (HMI).**

1.4. TARGET RESOLUTION

The **TARGET RESOLUTION** of a radar is its ability to distinguish between targets that are very close together in either range or bearing. Weapons-control radar, which requires great precision, should be able to distinguish between targets that are only yards apart. Search radar is usually less precise and only distinguishes between targets that are hundreds of yards or even miles apart.

Resolution is usually divided into two categories :

1. RANGE RESOLUTION and
2. BEARING RESOLUTION.

1.4.1. Range Resolution

Range resolution is the ability of a radar system to distinguish between two or more targets on the same bearing but at different ranges. The degree of range resolution depends on the width of the transmitted pulse, the types and sizes of targets, and the efficiency of the receiver and indicator. Pulse width is the primary factor in range resolution. A well-designed radar system, with all other factors at maximum efficiency, should be able to distinguish targets separated by one-half the pulse width time. Therefore, the theoretical range resolution of a radar system can be calculated from the following formula :

$$R_{RES} = c \times PW/2$$

The above formula is often written as :

$$R_{RES} = c/2\beta \qquad (PW = 1/\beta)$$

where β is the bandwidth of transmitter pulse.

For example, if a radar system has a pulse width of 5 microseconds, the range resolution is calculated as follows :

$$R_{RES} = c \times PW/2$$
$$R_{RES} = 3 \times 10^8 \times 5 \times 10^{-6}/2$$
$$= 3 \times 5 \times 10^2/2 = 7.5 \times 100$$
$$R_{RES} = 750 \text{ m}$$

In the above example, targets on the same bearing would have to be separated by more than 750 m to show up as two targets on your indicator.

1.4.2. Bearing Resolution

Bearing, or azimuth, resolution is the ability of a radar system to separate objects at the same range but at different bearings. The degree of bearing resolution depends on radar beam width and the range of the targets.

Range is a factor in bearing resolution because the radar beam spreads out as range increases.

A **RADAR BEAM** is defined in width in terms of HALF-POWER POINTS. All the points off the centerline of the beam that are at one-half the power level at the center are plotted to define beam width. When the half-power points are connected to the antenna by a curve, such as that shown in Fig. 1.10, the resulting angular width of the curve is called the **ANTENNA BEAM WIDTH**. The physical size and shape of the antenna determines beam width. Beam width can vary from about 1 degree up to 60 degrees. In Fig. 1.10, only the target within the half-power points will reflect a useful echo. Two targets at the same range must be separated by at least one beam width to be distinguished as two objects.

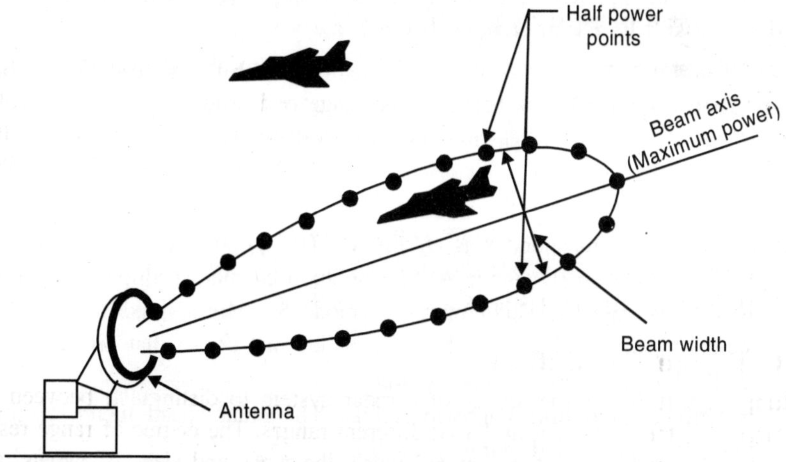

Fig. 1.10. Beam Half-power Points.

1.5. APPLICATIONS OF RADAR

Radar has been employed to detect target on the ground on the sea, in the air in space and even below ground. The major areas of radar applications are briefly described below :

1. Military applications. In military radar has wide application. Radar is used by the military specially for surveillance purpose of the enemy aircraft. The surveillance radat detect, locate and identified the enemy targets. It is also being used for providing navigational aids to military or as well as civil aircraft. It is also used for tracking the targets and controlling of weapons.

2. Targeting radars use the same principle but scan a much smaller area far more often, usually several times a second or more, where a search radar might scan a few times per minute. Some targeting radars have a range gate that can track a target, to eliminate clutter and electronic counter-measures.

3. Navigational radars resemble search radar, but use very short waves that reflect from earth and stone. They are common on commercial ships and long-distance commercial aircraft.

4. Weather radars can resemble search radars. These radar use radio waves with horizontal, dual (horizontal and vertical), or circular polarization. The frequency selection of weather radar is a performance compromise between precipitation reflectivity and attenuation due to atmospheric water vapour. Some weather radar uses doppler to measure wind speeds.

5. General purpose radars are increasingly being substituted for pure navigational radars. These generally use navigational radar frequencies, but modulate the pulse so the receiver can determine the type of surface of the reflector. The best general-purpose radars distinguish the rain of heavy storms, as well as land and vehicles. Some can superimpose sonar and map data from GPS position.

6. Radar Proximity Fuses are attached to anti-aircraft artillery shells or other explosive devices, and detonate the device when it approaches a large object. They use a small rapidly pulsing omnidirectional radar, usually with a powerful battery that has a long storage life, and a very short operational life. The fuses used in anti-aircraft artillery have to be mechanically designed to accept fifty thousand gravities of acceleration, yet still be cheap enough to throw away.

7. Radar altimeters measure an aircraft's true height above ground.

8. Air Traffic Control uses Primary and Secondary Radars

(*a*) *Primary radar* is a "classical" radar which reflects all kind of echoes, including aircraft and clouds.

(*b*) *Secondary radar* emit pulses and listen for special answer of digital data emitted by an Aircraft Transponder as an answer. Transponders emit different kind of data like a 4 octal ID (mode A), the onboard calculated altitude (mode C) or the Callsign (not the flight number) (mode S). Military use transponders to establish the nationality and intention of an aircraft, so that air defenses can identify possibly hostile radar returns.

9. Law Enforcement and Highway safety. The radar speed meter, familiar to many is used by police for enforcing speed limit.

10. Radar has also found application in industry for the non-contact measurement of speed and distance. It has been used for oil and gas exploration.

1.6. RADAR FREQUENCIES

Conventional radar generally operates in what is called the microwave region (a term not rigidly defined). Microwaves are since they are defined in terms of their wavelength (λ) in the sense that micro refers to tinyness. In short a microwave is a signal that has a wavelength of 1 foot or less *i.e.,* $\lambda \leq 30.5$ cms. So all the frequencies above approximately 1000 MHz (1 GHz) to about 1000 GHz are microwave frequencies. IEEE standard Radar Frequencies Letter–Band nomenclature given in Table 1.1.

EM spectrum frequency range and their general applications are given in Table 1.2.

Table 1.1. IEEE Standard Radar Frequencies

Band Designation	*Nominal Frequency range*	*Specific radar frequency range Based on ITU*
HF	3-30 MHz	
VHF	30-300 MHz	138-144 MHz
		216-225 MHz
UHF	300-1000 MHz	420 –450 MHz
		850-942 MHz
L	1-2 GHz	1215-1400 MHz
S	2-4 GHz	2300-2500 MHz
		2700-3700 MHz
C	4-8 GHz	5250-5925 MHz
X	8-12 GHz	8500-10680 MHz
KU	12-18 GHz	13.4-14 GHz
		15.7-17.7 GHz
K	18-27 GHz	24.05-24.25 GHz
Ka	27-40 GHz	33.4-36 GHz
V	40-75 GHz	59-64 GHz
W	75-110 GHZ	76-81 GHz
		92-100 GHz
mm	110-300 GHz	126-142 GHz
		144-149 GHz
		231-235 GHz
		238-248 GHz

Table 1.2. EM Spectrum Frequency Range and Their General Applications

Frequency	*Description*	*Application*
Upto 300 Hz	Extremely low frequency (ELF)	Special Communications
300 Hz – 3 kHz	Voice frequency	
3 kHz – 30 kHz	Very low frequency (VLF)	Shore-based communications, experimental
30 kHz – 300 kHz	Low frequency (LF)	Shore-based communications, navigation
300 kHz – 3 MHz	Medium frequency (MF)	Commercial broadcast band (550 kHz to 1700 kHz), communications on either side of the broadcast band
3 MHz – 30 MHz	High frequency (HF)	Ship and shore long-range communications
30 MHz – 300 MHz	Very high frequency (VHF)	Communications, navigation
300 MHz – 3 GHz	Ultra high frequency (UHF)	Line-of-sight communications to 400 MHz, above his frequency, radar and special equipments
3 GHz – 30 GHz	Superhigh frequency (SHF)	Radar and special equipments
30 GHz – 300 GHz	Extremely high frequency (EHF)	Radar and special equipments

1.7. RADAR DEVELOPMENT

Fig. 1.11. An Imagination Picture of Radar.

History of Radar

In 1887 the German physicist Heinrich Hertz began experimenting with radio waves in his laboratory. He found that radio waves could be transmitted through some materials, and were reflected by others. Although this was predicted earlier by James Clerk Maxwell, his work was not widely known as the time and much of the research became known through Hertz. In fact today we use his name as the SI unit of frequency, the hertz, often abbreviated to Hz.

Some years later a German engineer, Chistian Huelsmeyer, proposed the use of radio echoes to avoid collisions in marine navigation. In 1904 he completed construction of a device he called the **telemobiloscope**, which consisted of a simple spark gap aimed using a funnel-shaped metal antenna. When a reflection was seen by the two straight antennas attached to the receiver, a bell sounded. Although very simple, the system could detect shipping accurately up to about 3 km. Nevertheless the naval world seemed uninterested in his invention, and it was not put into production.

Invention of Modern Radar

The invention of modern radar is generally credited to Sir Robert Watson-Watt. In 1915 he joined the Royal Aircraft Factory at Ditton Park as a meteorologist, where he attempted to use radio signals generated by lightning strikes to map out thunderstorms. The difficulty in pinpointing the direction of these high-speed signals led to the use of rotating directional antennas, and in 1923 the use of oscilloscopes in order to display them in 2-D. At this point the only missing part of a functioning radar was the broadcaster.

Nikola Tesla, in August 1917, first established principles regarding frequency and power level for the first primitive RADAR units in 1934. In the 1917 *The Electrical Experimenter*, Tesla stated the principles of modern military radar in detail. Tesla's study of high voltage, high frequency alternating currents led to this development. Tesla had formed the concept of using radio waves to detect objects at a distance.

Tesla stated,

"For instance, by their [standing electromagnetic waves] use we may produce at will, from a sending station, an electrical effect in any particular region of the globe; [with which] we may determine the relative position or course of a moving object, such as a vessel at sea, the distance traversed by the same, or its speed."

Tesla proposed to use electromagnetic waves to determine the relative position, speed, and course of a moving object and other modern concepts of radar. Tesla had proposed that it might help find submarines (for which it isn't well-suited), though it was first applied successfully to find aircraft (after their later proliferation) and surface ships during World War II. Emil Girardeau, working with the first French radar systems, stated he was building radar systems "*conceived according to the principles stated by Tesla*".

In 1934, Watson-Watt was well established in the area of radio, and was approached by H.E. Wimperis from the Air Ministry, who asked about the use of radio to produce a 'death ray'. While he knew this to be unlikely, he did mention Meanwhile attention is being turned to the still difficult, but less unpromising, problem of radio detection and numerical considerations on the method of detection by reflected radio waves will be submitted when required. Watson-Watt and his assistant Arnold Wilkins published a report on the topic in February 1935, titled *The Detection of Aircraft by Radio Methods*.

Meanwhile in Germany, Hans Eric Hollmann had been working for some time in the field of microwaves, which were to later become the basis of almost all radar systems. In 1935 he published *Physics and Technique of Ultrashort Waves*, which was then picked up by researchers around the world. At the time he had been most interested in their use for communications, but he and his partner Hans-Karl von Willisen had also worked on radar-like systems.

In the autumn of 1934 their company, GEMA, built the first commercial radar system for detecting ships. Operating in the 50 cm range it could detect ships up to 10 km away, similar in purpose to Huelsmeyer's earlier device. In the summer of 1935 a pulse radar was developed with which they could spot the ship the *Königsberg* 8 km away, with an accuracy of up to 50 m, enough for gun-laying. The same system could also detect an aircraft at 500 m altitude at a distance of 28 km. The military implications were not lost this time around, and construction of land and sea-based versions took place as Freya and Seetakt.

Development in England and Germany

At this point both England and Germany knew of the ongoing efforts by their "competition". Both nations were intensely interested in the other's developments in the field, and engaged in an active campaign of espionage and false leaks about their respective equipment. But it was only in Britain that the usefulness of the system became obvious, so while the German systems had the edge technologically (operating on much shorter wavelengths) only Britain started true mass deployment of both the radars and the control systems needed to support them.

Invention of Chain Home in England

Shortly before the outbreak of World War II several radar stations known as **Chain Home** (or *CH*) were constructed in the south of England. As one might expect from the first radar to be deployed, CH was a simple system. The broadcast side was formed from two 300′ (100 m) tall steel towers strung with a series of cables between them.

The output of a powerful 50 MHz radio of about 200 kW (up to 800 kW in later models) was fed into these cables, pulsed at about 50 times a second. A second set of 240′ (73 m) tall wooden towers were used for reception, with a series of crossed antennas at various heights up to 215′ (65 m). In fact most stations had more than one set of each antenna, tuned to operate at different frequencies.

The CH radar was read with an oscilloscope. When a pulse was sent out into the broadcast towers, the scope was triggered to start its beam moving horizontally across the screen very rapidly. The output from the receiver was amplified and fed into the vertical axis of the scope, so a return from an aircraft would deflect the beam upward. This formed a spike on the display, and the distance from the left side-measured with a small scale on the bottom of the screen - would give the distance to the target. By rotating the receiver antennas to make the display disappear, the operator could determine the direction (this is the reason for the cross shaped antennas), the size of the vertical displacement indicated something of the number of aircraft involved, and by comparing the strengths returned from the various antennas up the tower, you could determine the height.

CH proved highly effective during the Battle of Britain, and are often credited with allowing the RAF to defeat the much larger Luftwaffe forces. Whereas the Luftwaffe had to hunt all over to find the RAF fighters, the RAF on the other hand knew exactly where the Luftwaffe bombers where, and could converge all of their fighters on them.

Very early in the battle the Luftwaffe made a series of small raids on a few of the stations, but they were eventually returned to operation in a few days. In the meantime the operators took to broadcasting radar-like signals from other systems in order to fool the Germans into believing that the systems were still operating. Eventually they gave up trying to bomb them. The Luftwaffe apparently never understood the importance of radar to the RAF's efforts, or they would have assigned them a much higher priority — it is clear they could have knocked them out continually if they wished.

In order to avoid the CH system the Luftwaffe adopted other tactics. One was to approach Britain at very low levels, below the sight line of the radar stations. This was countered to some degree with a series of shorter range stations built right on the coast, known as Chain Home Law (*CHL*). These radars had originally been intended to use for naval gun-laying as known as Coastal Defense (*CD*), but their narrow beams also meant they could sweep an area much closer to the ground without seeing the reflection of the ground (or water) itself. Unlike the larger CH systems, CHL had to have the broadcast antenna itself turned, as opposed to just the receiver. This was done manually on a pedal-crank system run by WAAFs until more reliable motorized movements were installed in 1941.

Later Adaptations

Similar systems were later adapted with a new display to produce the **Ground Controlled Intercept stations** starting in late 1941. In these systems the antenna was rotated mechanically, followed by the display on the operators console. That is, instead of a single line across the bottom of the display from left to right, the line was rotated around the screen at the same speed as the antenna was turning.

The result was a 2-D display of the air around the station with the operator in the middle, with all the aircraft appearing as dots in the proper location in space. These so-called **Plan Position Indicators** (*PPI*) dramatically simplified the amount of work needed to track a target on the operator's part. Such a system with a rotating, or sweeping, line is what most people continue to associate with a radar display.

Rather than avoid the radars, the Luftwaffe took to avoiding the fighters by flying at night and in bad weather. Although the RAF was aware of the location of the bombers, there was little they could do about them unless the fighter pilots could see the opposing planes. However, just this eventuality had already been forseen, and Watson-Watt (likely at the urging of Tizzard) had already started work on a miniaturized radar system suitable for aircraft, the so-called **AI** (*airborne interception*) set. Initial sets were available in 1941 and fitted to Bristol Blenheim aircraft, replaced quickly with the better performing Bristol Beaufighter, which quickly put an end to German night- and bad-weather bombing over England.

The next major development in the history of radar was the invention of the **cavity magnetron** by Randall and Boot of Birmingham University in early 1940. This was a small device which generated much more powerful microwaves than previous devices, which in turn allowed for the detection of much smaller objects and the use of much smaller antennas. The secrecy of the device was so high that it was decided in 1940 to move production to the USA, which resulted in the creation of the MIT Radiation Lab to develop the device further.

German Developments

German developments mirrored those in England, but it appears radar received a much lower priority until later in the war. The Freya was in fact much more sophisticated than its CH counterpart, and by operating in the 1.2 m wavelength (as opposed to ten times that for the CH) the Freya was able to be much smaller and yet offer better resolution. Yet by the start of the war only eight of these units were in operation, offering much less coverage.

However the Germans did not have an airborne system of any sort deployed until 1942, leaving them with the problem of having to get their fighters into that 300 m range solely with ground-based equipment. To fill this need another system known as Wuerzburg was deployed, starting in 1941.

Unlike other systems, the Wuerzburg was mounted on a highly directional parabolic antenna that was sensitive in only one direction. This made it useless for finding the targets, but once guided to one by an associated Freya it could track it with extreme accuracy: later models were accurate to 0.2 degrees or less. In order to do this the radar sent out two lobes and the return of each was shown on the display. By keeping the returns from both the same strength, the operator kept the Wuerzburg pointed directly at the target.

The downfall of the German radar network was that it could only track a single aircraft per Wuerzburg. In fact the system required two Wuerzburgs per interception, one for the target, and one for the fighter. This meant that as a raid developed, only a few night fighters could be directed at any one time, as only a small number of the eventual 5,000 Wuerzburgs would be within their 25 km range at any one time.

1.8. DIFFERENT TYPES OF RADAR

Radar systems may be divided into types based on the designed use. This section presents the general characteristics of several commonly used radar systems:

- Air-Defense Radars
- Air Traffic Control Radars

- Fire-control radar or tracking radar
- Speed gauges
- Airborne radar
- Mortar locating radar
- Radar Satellites
- Weather Radar
- Ground Penetrating radar

Depending the desired information radar units must have different qualities and technologies

On reason of these different qualities and techniques radar units are classified is:

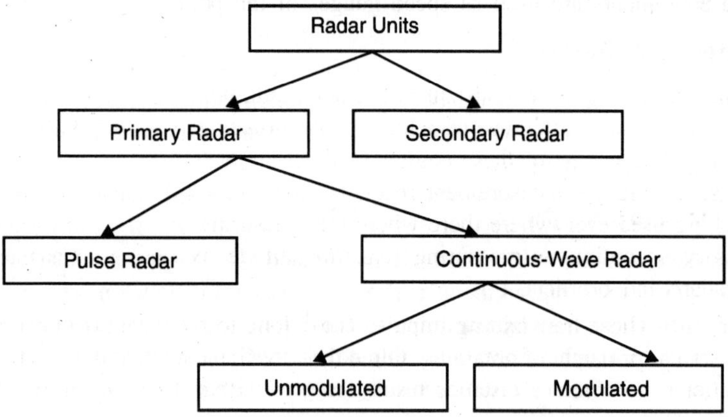

Fig. 1.12. Radar Classifications.

Primary Radar

A **primary radar** transmits high-frequency signals which are reflected at targets. The arisen echoes are received and evaluated. This means, unlike secondary radar units a primary radar unit receive its own emitted signals as an echo again.

Secondary Radar

At these radar units the airplane must have a transponder (transmitting responder) on board and receives an encoded signal of the secondary radar unit. An active also encoded response signal, which is returned to the radar unit then is generated in the transponder. In this response can be contained much more information, as a primary radar unit is able to acquire (E.g. an altitude, an identification code or also any technical problems on board such as a radio contact loss...). Example of secondary radar is IFF (Identification of Friend and Foe).

Pulse Radar

Pulse radar units transmit a high-frequency impulsive signal of high power. After this a longer break in which the echoes can be received follows before a new transmitted signal is sent out. Direction, distance and sometimes if necessary the altitude of the target can be determined from the measured antenna position and propagation time of the pulse-signal.

Continuous-Wave Radar (CW Radar)

CW radar units transmit a high-frequency signal continuously. The echo signal permanently is received and processed. The receiver needn't be mounted at the same place as the transmitter absolutely. Every firm civil radio transmitter can work as a radar transmitter at the same time, if a remote receiver compares the propagation times the direct one with the reflected signal. Tests are known that the correct location of an airplane can be calculated from the evaluation of the signals by three different television stations.

Unmodulated CW-Radar

The transmitted signal of these equipments is constant in amplitude and frequency. These equipments are specialized in speed measuring. Distances cannot be measured. E.g. these equipments are used as speed gauges of the police.

Modulated CW-Radar

The transmitted signal is constant in the amplitude but modulated in the frequency. This one gets possible after the principle of the propagation time measurement with that again. It is advantage of these equipments that an evaluation is carried out without reception break and the measurement result is therefore continuously available. These radar units are used everywhere there where the measuring distance isn't too large and it's necessary a continuous measuring (e.g. an altitude measuring in airplanes or as weather radar/wind profiler).

Radar units whose transmitting impulse is too long to get a well distance resolution also use a similar principle of obtaining. Often these equipments modulate its transmitting pulse additional, to obtain a distance resolution also within the transmitting pulse with help of the pulse-compression.

1.9. BASIC PULSE RADAR SYSTEM

A very basic radar block diagram of pulsed radar system was shown in Fig. 1.3. A more detailed block diagram of high–power pulsed radar set is shown in Fig. 1.13.

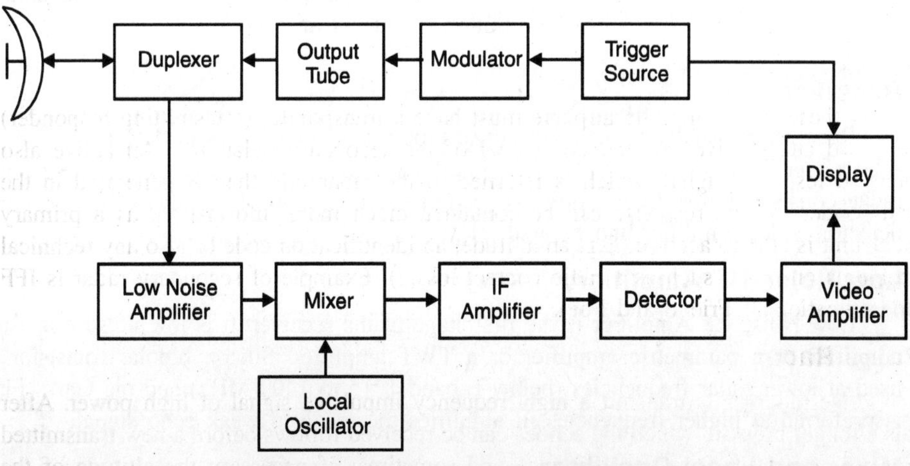

Fig. 1.13. Pulsed Radar Block Diagram.

Duplexer. This is a switch, which alternately connects the transmitter or receiver to the antenna. Its purpose is to protect the receiver from the high power output of the transmitter. During the transmission of an outgoing pulse, the duplexer will be aligned to the transmitter for the duration of the pulse, PW. After the pulse has been sent, the duplexer will align the antenna to the receiver. When the next pulse is sent, the duplexer will shift back to the transmitter. A duplexer is not required if the transmitted power is low.

Antenna. The antenna takes the radar pulse from the transmitter and puts it into the air. Furthermore, the antenna must focus the energy into a well-defined beam which increases the power and permits a determination of the direction of the target. The antenna must keep track of its own orientation, which can be accomplished by a synchro-transmitter. There are also antenna systems, which do not physically move but are steered electronically.

Transmitter (Output Tube). The transmitter creates the radio wave to be sent and modulates it to form the pulse train. The transmitter must also amplify the signal to a high power level to provide adequate range. The source of the carrier wave could be a Klystron, Travelling Wave Tube (TWT) or Magnetron. Each has its own characteristics and limitations.

Radar Modulator

The radar modulator is a device, which provides the high power to the transmitter tube to transmit during transmission period. It makes the transmitting tube ON and OFF to generate the desired waveform. Modulator allows storing the energy in a capacitor bank during the rest time.

The stored energy then can be put into the pulse when transmitted. It provides rectangular voltage pulses which acts as the supply voltage to the output tube such as magnetron, thus switching it ON and OFF as required. Commonly a thyratom valve is being used as switching device, which is a gas filled valve.

Trigger Source (Synchronizer)

The synchronizer coordinates the timing for range determination. It regulates that rate at which pulses are sent (*i.e.* sets PRF) and resets the timing clock for range determination for each pulse. Signals from the synchronizer are sent simultaneously to the transmitter, which sends a new pulse, and to the display, which resets the return sweep.

Receiver

Receiver is usually of superhetrodyne type whose function is to detect the desired echo signal in the presence of noise, interference and clutter. The receiver in pulsed radar consists of the Low Noise RF Amplifier, mixer, local oscillator IF amplifier, detector, video amplifier and radar display.

Low Noise RF Amplifier

Low Noise RF Amplifier is the first stage of the receiver. It is low noise transistor amplifier or a parametric amplifier or a TWT amplifier. Silicon bipolar transistor is used at lower radar frequencies (below L-band 1215 to 1400 MHz) and the GaAs FET is preferred at higher frequencies. It amplifies the received weak echo signal.

Mixer and Local Oscillator

These convert RF signal output from RF amplifier to comparatively lower frequency levels called Intermediate Frequency (IF). The typical value for pulse radar is 30 MHz or 60 MHz.

IF Amplifier

Its consists of a cascade of tuned amplifiers, these can be synchronous, that is, all tuned to the same frequency and having identical band pass characteristics. If a really large bandwidth is needed, the individual IF amplifiers may be stagger tuned. The typical value for pulse radar is 30 MHz or 60 MHz.

Detector

Detector is often a Schottky-barrier diode which extracts the pulse modulation from the IF amplifier output. The detector output is then amplified by the Video Amplifier to a level where it can be properly displayed usually on CRT (Cathode Ray Tubes) directly or via digital signal processors. Synchronizing pulses are applied by the trigger source to the display unit.

Display Unit

The received video signals are displays on the CRT (Cathode Rays Tube) for further observations and actions. Different types of Display system which are used in radar, discussed in Chapter 8 in this book.

SOLVED EXAMPLES

Example 1.1. Calculate the range of a target if the time taken by the radar signal to travel to the target and back is 100 μs.

Solution.

$$\text{Range } (R) = c \times \left(\frac{\Delta t}{2}\right)$$

$$c = 3 \times 10^8 \text{ m/s}$$

$$\Delta t = 100 \text{ μs}$$

By putting the values in equation, we get

$$= (3 \times 10^8) \times \left(\frac{100 \times 10^{-6}}{2}\right)$$

$$\boxed{R = 15 \text{ km.}} \quad \textbf{Ans.}$$

Example 1.2. If the transmitted peak power of radar is 100 kW, pulse repetition frequency is 1000 pps and the pulse width is 1 μs then calculate the average power in dbs.

Solution. We have

$$P_t = 100 \text{ kW}$$

$$PRF = 1000 \text{ pps}$$

$$PW = 1 \text{ μs}$$

We know that $\quad \text{Duty Cycle} = PW \times PRF$

By putting the value in the above equation, we get

$$\text{Duty cycle} = (10^{-6}) \times (1000) = 0.001$$

We know, The average power $= P_t \times$ Duty cycle

$$= (100 \times 10^3) \times 0.001$$

$$\boxed{\text{Average power} = 100 \text{ W}} \quad \textbf{Ans.}$$

or we can write in dbs, $\quad 100 \text{ W} = 20 \text{ dbs}$

Example 1.3. A typical pulse waveform of radar is shown below. In which some parameters of radar is shown. Calculate the (*a*) Average Power, (*b*) Duty Cycle and (*c*) Maximum range of radar.

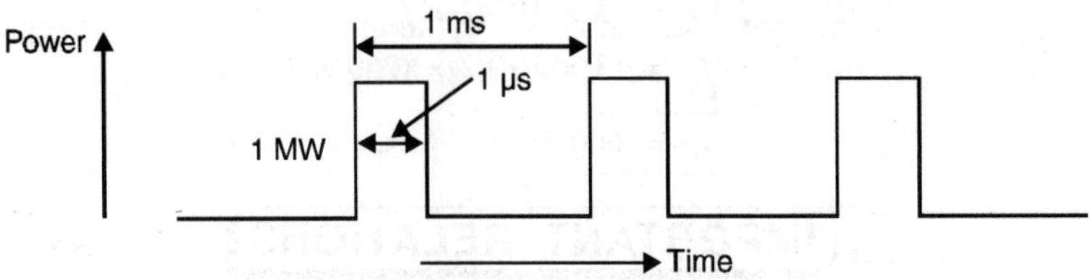

Fig. 1.14.

Solution. We have,

$$\text{Peak power } (P_t) = 1 \text{ MW}$$

$$\text{Pulse Width } (T) = 1 \ \mu\text{s}$$

$$\text{Pulse Repetition Period } (T_r) = 1 \text{ ms}$$

We know,

$$PRF\ (f_r) = 1/T_r = 1/(1 \times 10^{-3})$$

$$PRF = 1000 \text{ Hz}$$

(*a*)

$$P_{av} = P_t \times T \times f_r = (10^6) \times 1 \times 10^{-6} \times (1000)$$

$$\boxed{P_{av} = 1000 \text{ W} = 1 \text{ kW}} \quad \textbf{Ans.}$$

(*b*)

$$\text{Duty Cycle} = T \times f_r = (10^{-6}) \times (1000) = 0.001$$

$$\boxed{\text{Duty Cycle} = 0.001} \quad \textbf{Ans.}$$

(*c*) Maximum Unambiguous Range (R_{max}) = $c/(2 \times f_r)$

$$= \frac{(3 \times 10^8)}{(2 \times 1000)} = (3/2) \times 10^5$$

$$\boxed{R_{max} = 150 \text{ km}} \quad \textbf{Ans.}$$

or

$$R_{max} = \frac{150 \times 10^3}{1852} = 81 \text{ nmi} \qquad (1 \text{ nmi} = 1853 \text{ m})$$

Example 1.4. Determine the maximum unambiguous range and range resolution of a pulse radar having pulse width is 5 µs at a PRF of 1000 Hz.

Solution.

$$R_{max} \text{ (unamb)} = c/(2 \times f_r) = \frac{(3 \times 10^8)}{(2 \times 1000)} = 150 \text{ km}$$

$$\boxed{R_{max} \text{ (unamb)} = 150 \text{ km}} \quad \textbf{Ans.}$$

$$\text{Range Resolution} = c \times \frac{T}{2} = \frac{(3 \times 10^8) \times (5 \times 10^{-6})}{2} = 7.5 \times 100 = 750 \text{ m}$$

$$\boxed{\text{Range Resolution} = 750 \text{ m}} \quad \textbf{Ans.}$$

Example 1.5. A radar is to have a maximum range of 250 km. Determine the maximum allowable PRF for unambiguous reception.

Solution.

$$R_{max} \text{ (unamb)} = 250 \text{ km} = c/(2 \times f_r)$$

$$(250 \times 10^3) = (3 \times 10^8)/2 \times f_r$$

$$f_r = (3 \times 10^8)/2 \times (250 \times 10^3)$$

$$\boxed{f_r = 600 \text{ Hz}} \quad \textbf{Ans.}$$

IMPORTANT RELATIONS

1. $R = C \times \dfrac{\Delta t}{2}$
2. $PRT = PW + RT$
3. $PRF = \dfrac{1}{PRT}$
4. $P_{av} = P_t \times \dfrac{PW}{PRT} = P_t \times PW \times PRF$
5. Average Power = Peak Power × $\dfrac{\text{Pulse Width}}{\text{Pulse Repetition Time}}$

   = Peak Power × Pulse Width × Pulse Repetition Frequency
6. Duty cycle = $PW \times PRF$
7. $R_{RES} = C \times \dfrac{PW}{2}$
8. $R_{RES} = \dfrac{C}{2} \times \dfrac{1}{\beta}$ $\left(\beta = \text{Bandwidth of transmitter} \times \text{Pulse; } PW = R_{RES} = \dfrac{1}{\beta}\right)$

where, R = Range of the target

C = Speed of light or propagation speed of EM wave = 3×10^8 m/s

Δt = Delay time between the transmission of radio energy and the return of reflected energy or round trip propagation time.

PRT = Pulse Repetition Time or Period

PRF = Pulse Repetition Frequency

P_{av} = Average Power

R_{RES} = Range Resolution

β = Bandwidth of Transmitter Pulse

SUMMARY

- Radar is an acronym for Radio Detection and Ranging.
- Echo is a small portion of the energy is reflected and returned to the radar.
- Seven Basic components of simple radar system.
  - (*a*) Transmitter used to generate the high power RF pulses to send to the space through antenna.
  - (*b*) Receiver used to received the echo signals from the target and process them.

(*c*) Power supply provided the electrical power for all the components. Transmitter needed maximum electrical power, which may be several kW.

(*d*) Synchroniser coorrdinates the timing for range determination.

(*e*) Duplexer act as a switch that alternately connects the transmitter or receiver to the antenna.

(*f*) Antenna radiate the *RF* energy into the space and received the reflected signal from the target.

(g) Display unit used to represent the target location, in range, azimuth and elevation as per the radar applications.

- Range determination is accomplished by measuring the time taken by *RF* energy EM wave to travel from radar to target and back. It can be determined by using following formula, $R = c \times \Delta t/2$ where $c = 3 \times 10^8$ m/s.
- Common parameters of Radar pulse :

(*a*) *PW* is the duration for which radar transmitter is ON.

(*b*) Rest time (*RT*) is the interval between the pulses and to receive echo signal from the target, during the silence period.

(*c*) *PRT* is the interval between the start of one pulse and the start of another. It may written as $PRT = PW + RT$.

(*d*) *PRF* is the number of pulses transmitted in one second and is equal to inverse of *PRT*. $PRF = 1/PRF$.

(*e*) P_t (Peak power) is defined as the power averaged over that carrier frequency cycle which occurs at the maximum of the pulse power.

(*f*) Average Power (P_{av}) is defined as the average transmitted power over the pulse repetition period. It my be written as

$$P_{av} = P_t \times \frac{PW}{PRT} = P_t \times PW \times PRT$$

(*g*) Duty cycle is the ratio of average power to the peak power or *PW* to the *PRT* is called the duty cycle. It may be defined as the energy transmitted in one second by the transmitter is also as duty cycle.

$$\text{Duty cycle} = PW \times PRF$$

or

$$\text{Duty cycle} = \frac{P_{av}}{P_t}$$

- *Applications of Radar* :

(*a*) Surveillance search radar in military.

(*b*) Navigational radar in military as well as in civil.

(*c*) Targetting radar or tracking radar.

(*d*) Weather radar.

(*e*) Radar proximity fuses.

(*f*) Radar altimeter.

(*g*) Air traffic control radar.

(*h*) Traffic radar.

(*i*) Monitoring of space vehicle.

(*j*) In industry for oil and gas exploration.

- Radar generally operates in microwave region. The wavelength $\lambda \leq 30.5$ cm.
- History of radar.
- Different types of radar such as
  - (*a*) Air defence radar
  - (*b*) Air traffic control radar
  - (*c*) Fire control radar or tracking radar
  - (*d*) Airborne radar
  - (*e*) Mortar locating radar
  - (*f*) Ground penetrating radar.

UNSOLVED PROBLEMS

1. For a maximum unambiguous range of 1000 km, calculate the PRF required for the radar. [1500 Hz]
2. Determine the maximum unamb. range and range resolution of a pulse radar whose pulse duration is 10 μs at a PRF 1000 Hz. [150 km, 1.5 km]
3. A radar receiver having BW of 3 MHz, calculate the highest resolution in range. [50 m]
4. A pulse radar measure the round trip propagation time with an error of 2 μs. What will be the corresponding error in the measurement of range. [300 m]
5. Calculate the duty cycle of a radar which transmits a 1.5 ms pulse at a PRF of 8 kHz. If the peak power of this radar is 500 kilowatts, what is the average power ? What is the resting time ? [2×10^{-3}, 6 kW]
6. A pulsed radar has a duty cycle of 0.016. If the resting time is 380 ms, what is the pulse width ? What is the PRF ? What is the minimum range, in meters, of this radar ? [6.18 μs]
7. If a CRT is designed so that the spot traverses the indicator face in 900 ms, what range could be indicated by a spot deflection half way across the indicator ? [67 km]
8. What is the minimum range for radar with a PW of 5 ms ? [750 m]
9. What should be the PRF of radar in order to achieve a maximum unambiguous range of 50 nmi ? [1.61 kHz]
10. A radar transmitter has a peak power of 400 kW, a PRF of 1500 PPS and pulse width of 0.8 μs. Calculate : (*a*) maximum unamb. range; (*b*) duty cycle; (*c*) average power; (*d*) suitable bandwidth. [(*a*) 100 km (*b*) 0.0012 (*c*) 480 W (*d*) 1.25 MHz]

REVIEW QUESTIONS

1. What do you understand by terms RADAR and explain the echo pulse principle ?
2. What are the basic functions of radar ? Indicating the position of target, what is the difference between azimuth and elevation ?
3. Explain the working of pulse radar with the help of block diagram.
4. Explain the various applications of radar.
5. What are the radar frequencies, explain in brief ?
6. What is the maximum unambiguous range ? How it is related with pulse repetition rate?

7. Explain a Radar system with diagram. **[UPTU, 2003]**
8. Discuss Radar applications. **[UPTU, 2003]**
9. Explain the principle of radar system. **[UPTU, 2004]**
10. Explain the limitations and application of radars.
11. What is Radar ? Describe the principle operation of a basic radar with the help of neat block diagram. What are the various applications of Radar ?
12. What is the difference between pulse interval and PRF ?
13. What are the salient factors governing the choice of the PRF for a particular radar ? What are difference between the pulse interval and the PRF ?
14. Draw block diagram of a pulse radar and explain each block briefly. **[UPTU, 2006]**
15. Explain the block diagram of radar. Also discuss the radar frequencies according IEEE standard. Also discuss its applications.
16. Define the following terms in brief :
    - (*i*) Pulse Width
    - (*ii*) Pulse Repetition Time
    - (*iii*) Average Power
    - (*iv*) Duty Cycle
    - (*v*) Range Resolution
    - (*vi*) Azimuth Resolution
    - (*vii*) Radar frequencies
17. Draw the block diagram of a basic Radar set and explain the essentials of its operations. What are the basic functions of Radar ?

❑❑❑

CHAPTER

2

THE RADAR EQUATION

2.1. INTRODUCTION

INSIDE THIS CHAPTER

The radar equation relates the range of radar to the characteristics of the T_x (transmitter), R_x (receiver), A_e (antenna), target and the environment. It is useful to determine the maximum range at which particular radar can detect a target and its makes to understand the factor affecting radar performance. In this section, the simple form of equation is derived.

The radar equation is an important tool for following aspect:

1. Assessing the performance of radar.
2. Designing of new radar systems.
3. Assessing the technical requirement for new radar procurement.

2.2. SIMPLE FORM OF RADAR EQUATION

If the power of radar T_x is denoted by P_t and if an isotropic antenna (which will radiate energy uniformly in all direction) then the power density at a distance "R" from the radar is equal to the radiated power divided by the surface area $4\pi R^2$ of an imaginary sphere of radius R, or

i.e., power density at range R from an isotropic antenna

$$= P_t/4\pi R^2 \text{ watt/m}^2 \qquad ...(2.1)$$

However, we employ directive antenna (with narrow beam width) to concentrate the radiated power P_t in a particular direction.

The gain of an antenna is a measure of the increased power density radiated in some direction as compared to the power density that would appear in that direction from an isotropic antenna. The maximum gain "G" of an antenna may be defined as

$$G = \frac{\text{Maximum power density radiated by a directive antenna}}{\text{Power density radiated by loss less isotropic antenna}}$$

Power density at a distance R from directive antenna of power Gain "G",

$$= P_t\, G/4\pi R^2 \text{ watt/m}^2 \quad ...(2.2)$$

The target intercepts a portion of the incident energy and re-radiates it in various directions. It is only the power density re-radiated in the direction of the radar (echo signal) that is of interest. The radar cross-section of the target determines the power density returned to the radar for a particular power density incident on the target. It is denoted by 'σ' and is often called, for short, target cross-section or radar cross-section. The radar cross-section is defined by the following :

$$\text{Reradiated power density back at the radar} = \frac{P_t.G.\sigma}{4\pi R^2.4\pi R^2} \quad ...(2.3)$$

The radar cross-section has units of area, but it can be misleading to associate the radar cross-section directly with the targets physical size. Radar cross-section more dependent on the targets shape than on its physical size.

The radar antenna received a portion of the echo power. If the effective area of receiving antenna is denoted A_e, the power P_r received by the radar is,

$$P_r = \frac{P_t.G.\sigma.A_e}{4\pi R^2.4\pi R^2} \text{ watts} \quad ...(2.4)$$

The maximum range of radar R_{max} is the distance beyond which the target can not detected. It occurs when the received signal power P_r just equals the minimum detectable signal (S_{min}).

Substituting, $S_{min} = P_r$ in Eqn. (2.4)

$$S_{min} = \frac{P_t.G.\sigma.A_e}{4\pi R^2.4\pi R^2} \text{ watts} \quad ...(2.5)$$

i.e.,

$$R^4_{max} = \frac{P_t.G.\sigma.A_e}{(4\pi).(4\pi)(S_{min})} \quad ...(2.6)$$

$$R^4_{max} = \frac{P_t.G.\sigma.A_e}{(4\pi)^2.(S_{min})} \quad ...(2.7)$$

$$\boxed{R_{max} = \sqrt[4]{\frac{P_t.G.\sigma.A_e}{(4\pi)^2.S_{min}}}} \quad ...(2.8)$$

where,
P_t = Transmitter Power,
G = Maximum Gain of Antenna
Ae = Aperture area of receiving Antenna
σ = Cross section area of the target
S_{min} = Minimum detectable signal (mds)
R_{max} = Maximum range of radar

This is the fundamental form of radar range equation. The important parameters are the transmitting gain and the receiving effective area. The transmitter power P_t has not been specified as either the average or the peak power.

If the same antenna is used for both transmitting and receiving, as it usually is in radar, antenna theory gives the relationship between transmit gain G and the receive effective area A_e as

$$G = \frac{4\pi . A_e}{\lambda^2} \quad ...(2.9)$$

where λ = wavelength ($\lambda = c/f$ where c = velocity of propagation and f = frequency) Eq. (2.8) can be substituted, first for A_e and the G, to give two other forms of the radar equation,

$$R_{max} = \sqrt[4]{\frac{P_t . G^2 \sigma . \lambda^2}{(4\pi)^3 \; S_{min}}} \quad ...(2.10)$$

or

$$R_{max} = \sqrt[4]{\frac{P_t . A_e^2 \sigma}{4\pi\lambda^2 \; S_{min}}} \quad ...(2.11)$$

These three forms of the radar Eqs, (2.8), (2.10) and (2.11) are basically the same.

2.3. PREDICTION OF RANGE PERFORMANCE

The simple form of radar equation is not adequately predicting the range performance of actual radars. There are some drawbacks of simple form of radar equation appended below:

1. The fluctuation and uncertainties in targets radar cross-section.
2. The nature of minimum detectable signal, which is affected by the receiver noise.
3. The losses experienced in the radar system.
4. Propagation effect by earth surface and atmosphere.

The range of radar, therefore, will be a function of the Probability of detection (P_d) and the Probability of false alarm (P_{fa}).

2.3.1. Radar Performance

All of the parameters of the basic pulsed radar system will affect the performance in some way. Here we find specific examples and quantify this dependence where possible.

2.3.1.1. Transmitter Power

The power P_t in the radar equation is called by the radar engineer, the Peak Power. The peak power as used in the radar equation is not the instantaneous peak power of a sine wave. It is defined as the power averaged over that carrier frequency cycle which occurs at the maximum of the pulse power. Peak power is usually equal to one-half of the maximum instantaneous power.

The average power P_{av} is defined as the average transmitted power over the pulse repetition period. If the transmitted waveform is a train of rectangular pulses of width (T) and *PRP* (T_r).

$$\mathbf{PRP = 1/PRF} \quad ...(2.12)$$

$$\mathbf{P_{av} \text{ (Average Power)} = P_t \times T/T_r = P_t \times T \times PRF} \quad ...(2.13)$$

(Average Power = Peak Power × Pulse Width/*PRP* = Peak Power × Pulse width × PRF).

The ratio of average power to the Peak power or pulse width to the *PRP* or Pulse width × *PRF*, is called the **Duty Cycle** of the radar.

The fraction of time the radar transmitter (T_x) is generating pulse power is termed the Duty-Cycle. Thus, in any one second, a radar producing 1 μ sec pulses at a rate of 1000 per sec is "on" for

$$(1 \times 10^{-6}) \times (1000) = 0.001 \text{ s}$$

The duty cycle is then 0.001, and the average power is given by the product of duty cycle and peak power; thus if the peak power is 500 kW, the average power should be 500 W.

Typical values of radar power lies within the range 100 to 500 kW.

Writing the equation in terms of the average power rather than the peak power, we get

$$R^4_{max} = \frac{P_{av}.G.\sigma.A_e}{(4\pi)^2 S_{min} T.f_r} \quad ...(2.14)$$

where $f_r = PRF$

P_{av} = Average power

If the radar range is to be doubled, we have to increase the transmitter power 16 times since $R_{max} \propto$ ¼ root of peak power.

2.3.1.2. Pulse Width (PW)

The duration of the pulse and the length of the target along the radial direction determine the duration of the returned pulse. In most cases the length of the return is usually very similar to the transmitted pulse. In the display unit, the pulse (in time) will be converted into a pulse in distance. The range of values from the leading edge to the trailing edge will create some uncertainty in the range to the target. Taken at face value, the ability to accurately measure range is determined by the pulse width.

If we designate the uncertainty in measured range as the range resolution, R_{RES}, then it must be equal to the range equivalent of the pulse width, namely:

$$R_{RES} = c\ PW/2 \quad ...(2.15)$$

Now, you may wonder why not just take the leading edge of the pulse as the range which can be determined with much finer accuracy ? The problem is that it is virtually impossible to create the perfect leading edge. In practice, the ideal pulse will really appear like:

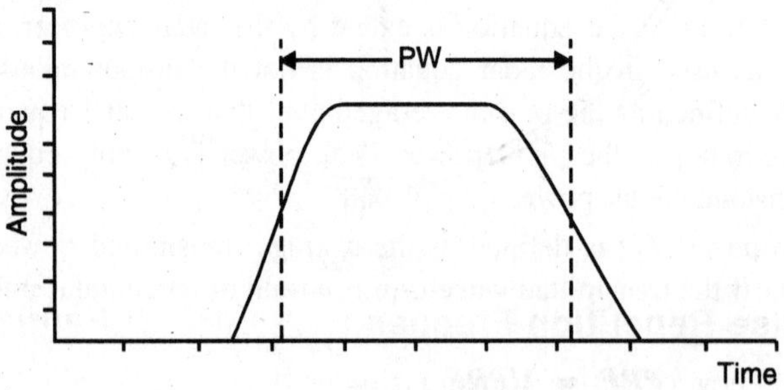

Fig. 2.1.

To create a perfectly formed pulse with a vertical leading edge would require an infinite bandwidth. In fact you may equate the Bandwidth, β, of the transmitter to the minimum pulse width, *PW* by :

$$PW = 1/\beta \quad ...(2.16)$$

Given this insight, it is quite reasonable to say that the range can be determined no more accurately than $cPW/2$ or equivalently

$$\boxed{R_{RES} = c/2\beta} \qquad ...(2.17)$$

In fact, high-resolution radar is often referred to as wide-band radar, which you now see as equivalent statements. One term is referring to the time domain and the other the frequency domain. The duration of the pulse also affects the minimum range at which the radar system can detect. The outgoing pulse must physically clear the antenna before the return can be processed. Since this lasts for a time interval equal to the pulse width, *PW*, the minimum displayed range is then:

$$\boxed{R_{MIN} = c \times \frac{PW}{2}} \qquad ...(2.18)$$

The minimum range effect can be seen on a *PPI* display as a saturated or blank area around the origin.

Increasing the pulse width while maintaining the other parameters the same will also affect the duty cycle and therefore the average power. For many systems, it is desirable to keep the average power fixed. Then the *PRF* must be simultaneously changed with *PW* in order to keep the product *PW* × *PRF* the same. For example, if the pulse width is reduced by a factor of ½ in order to improve the resolution, then the *PRF* is usually doubled.

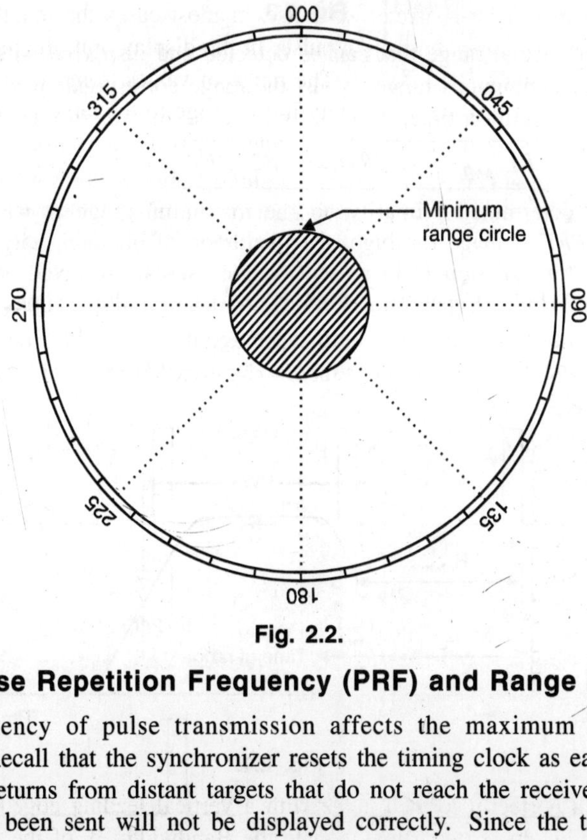

Fig. 2.2.

2.3.1.3. Pulse Repetition Frequency (PRF) and Range Ambiguities

The frequency of pulse transmission affects the maximum range that can be displayed. Recall that the synchronizer resets the timing clock as each new pulse is transmitted. Returns from distant targets that do not reach the receiver until after the next pulse has been sent will not be displayed correctly. Since the timing clock has been reset, they will be displayed as if the range where less than actual. If this were possible, then the range information would be considered ambiguous. An operator would not know whether the range were the actual range or some greater value.

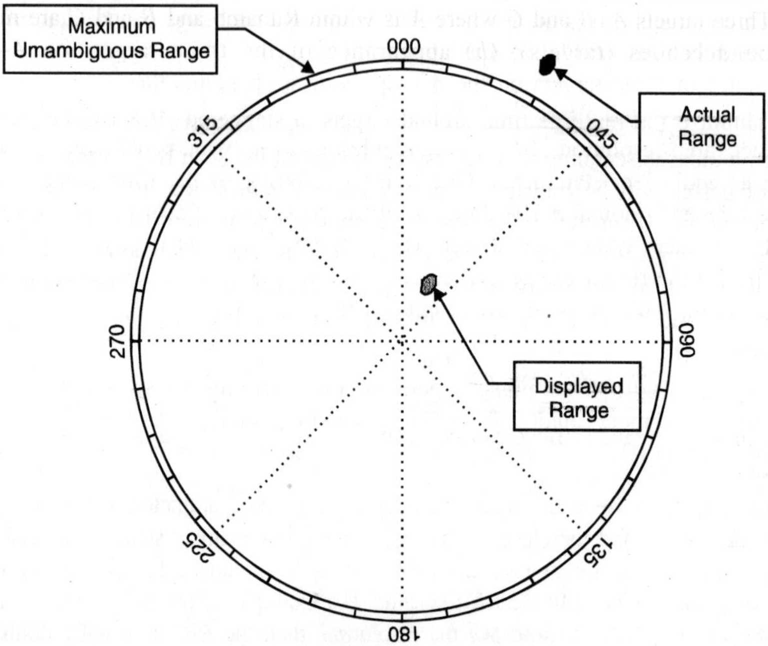

Fig. 2.3.

The maximum actual range that can be detected and displayed without ambiguity, or the maximum unambiguous range, is just the range corresponding to a time interval equal to the pulse repetition time, *PRT*. Therefore, the maximum unambiguous range,

$$\boxed{R_{UNAMB} = c\ PRT/2 = c/(2PRF)} \quad ...(2.19)$$

The *PRF* is determined primarily by the maximum range at which targets are expected. If the *PRF* is made too high, the likelihood of obtaining target echoes from the wrong pulse transmission is increased. Echoes signals received after an interval exceeding the *PRP* (Pulse Repetitation Period) are called multiple times around echoes.

They can result in erroneous or confusing range measurement. The nature of some multiple-time-around echoes may cause them to be labelled as ghost or angel targets or even flying saucers.

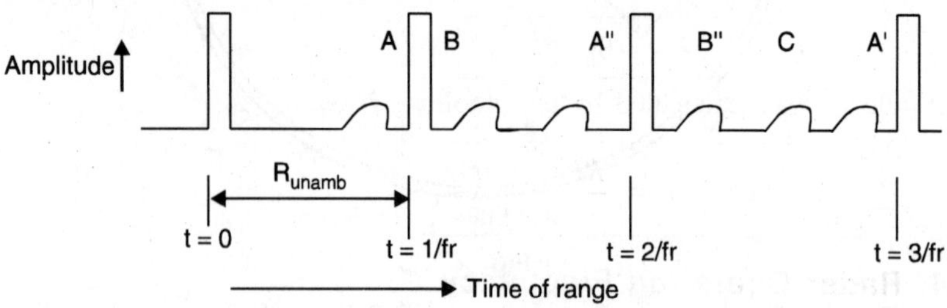

(*a*)

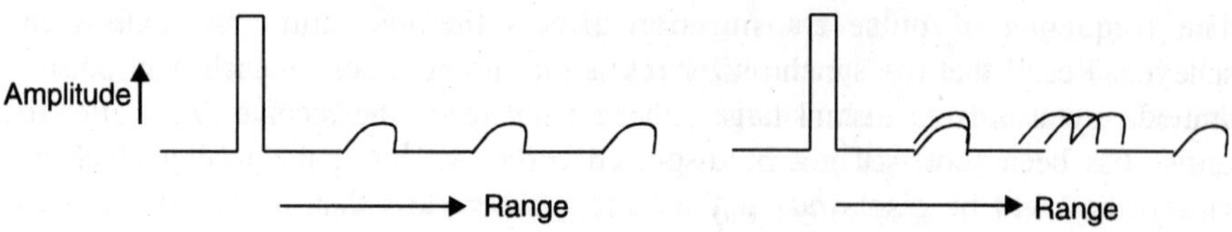

(*b*) (*c*)

Fig. 2.4 (*a*, *b* and *c*)

(*a*) Three targets *A, B* and *C* where *A* is within Runamb and *B* and *C* are multiple time around echoes (targets); (*b*) appearance of the three targets on *A*–scope (*c*) appearance of three targets on the *A*-scope with a changing *PRF.*

To minimize the multiple time around targets, a staggered *PRF* can be used.

When a radar is scanning, it is necessary to control the scan rate so that a sufficient number of pulses will be transmitted in any particular direction in order to guarantee reliable detection. If too few pulses are used, then it will more difficult to distinguish false targets from actual ones. False targets may be present in one or two pulses but certainly not in ten or twenty in a row. Therefore to maintain a low false detection rate, the number of pulses transmitted in each direction should be kept high, usually above ten.

For systems with high pulse repetition rates (frequencies), the radar beam can be repositioned more rapidly and therefore scan more quickly. Conversely, if the *PRF* is lowered the scan rate needs to be reduced. For simple scans it is easy to quantify the number of pulses that will be returned from any particular target. Let τ represent the dwell time, which is the duration that the target remains in the radar's beam during each scan. The number of pulses, *N*, that the target will be exposed to during the dwell time is:

$$\boxed{N = \tau PRF} \qquad ...(2.20)$$

We may rearrange this equation to make a requirement on the dwell time for a particular scan

$$\boxed{\tau_{min} = N_{min}/PRF} \qquad ...(2.21)$$

So it is easy to see that high pulse repetition rates require smaller dwell times. For a continuous circular scan, for example, the dwell time is related to the rotation rate and the beam-width.

$$\boxed{\tau = \theta_B / \theta_s} \qquad ...(2.22)$$

where θ_B = beam-width [degrees] θ_s = rotation rate [degrees/sec] which will give the dwell time (τ) in seconds. These relationships can be combined; giving the following equation from which the maximum scan rate may be determined for a minimum number of pulses per scan:

$$\boxed{\theta_s = \theta_B \; PRF/N} \qquad ...(2.23)$$

Scan rate may be calculated by using following equation,

$$\boxed{\text{Scan rate } (\theta_s) = \frac{RPM \times 360°}{60 \text{ s}}} \qquad ...(2.24)$$

2.3.1.4. Radar Operating Frequency

Finally, the frequency of the radio carrier wave will also have some affect on how the radar beam propagates. At the low frequency extremes, radar beams will refract in the atmosphere and can be caught in "ducts" which result in long ranges. At the high extreme, the radar beam will behave much like visible light and travel in very straight lines. Very high frequency radar beams will suffer high losses and are not suitable for long range systems.

The frequency will also affect the beam-width. For the same antenna size, low frequency radar will have a larger beam-width than a high frequency one. In order to keep the beam-width constant, low frequency radar will need a large antenna.

2.4. MINIMUM DETECTABLE SIGNAL

The receiver is sensitive to the range of frequencies being transmitted and provides amplification of the returned signal. In order to provide the greatest range, the receiver must be very sensitive without introducing excessive noise. The ability to discern a received signal from background noise depends on the signal-to-noise ratio (*S/N*).

The background noise is specified by an average value, called the noise-equivalent-power (*NEP*). This directly equates the noise to a detected power level so that it may be compared to the return. Using these definitions, the criterion for successful detection of a target is

$$P_r > (S/N)\ NEP,$$

where P_r is the power of the return signal. Since this is a significant quantity in determining radar system performance, it is given a unique designation, S_{min}, and is called the **Minimum Signal for Detection.**

$$S_{min} = (S/N)\ NEP \quad ...(2.25)$$

Since S_{min}, expressed in Watts, is usually a small number, it has proven useful to define the decibel equivalent, *MDS*, which stands for **Minimum Discernible Signal.**

$$MDS = 10 \text{ Log } (S_{min}/1 \text{ mW}) \quad ...(2.26)$$

When using decibels, the quantity inside the brackets of the logarithm must be a number without units. In the definition of MDS, this number is the fraction $S_{min}/1$ mW. As a reminder, we use the special notation dBm for the units of *MDS*, where the "*m*" stands for 1 mW. This is shorthand for decibels referenced to 1 mW, which is sometimes written as db/1 mW.

In the receiver, *S/N* sets a threshold for detection which determines what will be displayed and what will not. In theory, if *S/N* = 1, then only returns with power equal to or greater than the background noise will be displayed.

Detection is based on establishing a threshold level at the output of the receiver.

2.4.1. Threshold Detection

The target is said to be present when the receiver output crosses the threshold level and if the level of the receiver output signal is low and not able to cross the threshold level, it may be considered as noise. So the detection of target or noise may be decided by the crossing of threshold level and this type of detection is known as **threshold detection.**

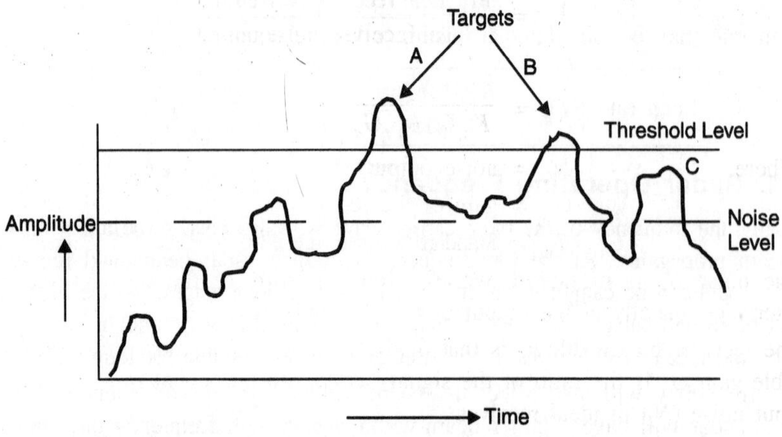

Fig. 2.5. Typical Output of the Radar as a Function of Time.

In the figure *A*, *B* and *C* represent signal pulse noise. *A* and *B* are valid detection, but *C* is missed detection. If the threshold level is set too low, noise might exceed it and be mistaken as target. This is called **False Alarm**. If the threshold were set too high, noise might not be large enough to cause false alarm, but weak targets echoes might not exceeds the threshold and would not be detected, when this occurs it is called **Missed Detection.**

2.5. RADAR RECEIVER NOISE AND SIGNAL-TO-NOISE RATIO

"**Noise** is unwanted Electromagnetic energy which interfere with the ability of the receiver to detect the wanted signal". Noise may be generated within the receiver it self or it may enter through the receiving antenna along with the reflected signals.

Thermal Noise or Johnson Noise is the noise, which is generated by the thermal motion of the conducted electrons in receiver input stages. The magnitude of the noise is directly proportional to the bandwidth and the absolute temperature of the resistive input circuit. The thermal noise power of a radar receiver may be written as

Thermal noise power (watt) $\propto T \,.\, B_n$

Or $= k.T \,.\, B_n$ ...(2.27)

Where, K = Boltzman's Constant = 1.38×10^{-23} Joules/degree

of a in Superhetrodyne receiver the bandwidth of the IF Amplifier which may be called as **matched filter.**

We can defined bandwidth noise (B_n) as

$$B_n = \frac{\int_0^{\infty} |H(f)|^2 \, dt}{|H(f_0)|^2} \quad ...(2.28)$$

In the above Eqn. (2.28),

f_0 = Frequency of the maximum response

$H(f)$ = The IF amplifier function of frequency response

The Noise Figure (F_n) in practical receiver may be defined as the ratio of noise out of a practical receiver to that noise out of an ideal receiver assuming that no other noise is present except thermal noise in the receiver.

Noise figure (F_n) may be written as

$$F_n = \frac{\text{Practical receiver Noise output}}{\text{Ideal receiver Noise output}}$$

$$F_n = \frac{N_0}{K.T_0.B_n.G_a} \quad ...(2.29)$$

Where, N_0 = noise output

G_a = gain of the receiver

T_0 = Standard temperature is to be taken 290 degree Kelvin.

The noise N_0 is measured over the linear portion of the receiver input-output characteristics, usually at the output of the IF amplifier.

The receiver bandwidth B_n is that of the IF amplifier in most radar receiver. The Available gain G_a is the ratio of the signal out (S_0) to the signal in (S_i), and kT_0B_n is the input noise (N_i) in ideal receiver.

$$F_n = \frac{S_i/N_i}{S_0/N_0} \quad \text{...(2.30)}$$

$$= \left(\frac{S_i}{S_0}\right) \times \left(\frac{N_0}{N_i}\right)$$

$$= \frac{S_i}{(G_a . S_i)} \times \left[G_a \times \frac{(N_i + N_r)}{N_i} \right]$$

$$F_n = \left[\frac{1 + N_{ri}}{N_i}\right]$$

Thus,
$$\left(\frac{N_{ri}}{N_i}\right) = (F_n - 1)$$

$$N_{ri} = N_i(F_n - 1)$$

Or
$$\boxed{N_{ri} = k \,.\, T_0 \,.\, B(F - 1) = S_{(\min)}}$$

$$\boxed{k \,.\, T_0 \,.\, B(F - 1) = S_{(\min)}}$$

Where N_{ri} = Noise power generated at the receiver input.

Substitute the value of $S_{(\min)}$ in Eqn. (2.11),

$$\boxed{R_{\max} = \sqrt[4]{\frac{P_t . A_e^2 . \sigma}{4\pi\lambda^2 . k . T_0 . B . (F-1)}}} \quad \text{...(2.31)}$$

By substituting the capture in terms of antenna diameter and for the various constants. From antenna theory, we know that

$$\boxed{A_e = \frac{0.65\pi(D^2)}{4}} \quad \text{...(2.32)}$$

We may further simplify the radar range equation and may be written as

$$\boxed{R_{\max} = (48)\sqrt[4]{\frac{P_t . D^4 . \sigma}{\lambda^2 . B . (F-1)}} \text{ Km}} \quad \text{...(2.33)}$$

or
$$R_{\max} = 48\left[\frac{(P_t . D^4 . \sigma)}{\lambda^2 B(F-1)}\right]^{1/4} \quad \text{...(2.34)}$$

or
$$\boxed{R_{\max} = 26\left[\frac{(P_t . D^4 . \sigma)}{\lambda^2 B(F-1)}\right]^{1/4} \text{ Nm}} \quad \text{(since 1 Nm = 1853 m)}$$

where, $R_{\max}$ = Maximum radar range,

P_t = Peak power

D = Antenna diameter in meters

σ = Radar cross-section of the target (m^2)

λ = wavelength in meter

F = Noise figure in ratio

2.6. SYSTEM LOSSES

In the radar equation one of important features has not being considered that is losses occurred in the radar system it self such as,

(*a*) The antenna beam shape loss,

(*b*) The collapsing loss,

(*c*) The losses in the microwave plumbing etc.

The losses are very significant while prediction of radar performance. In a complete radar system, there are so many losses, which are very negligible as an individual but if combine them that may be more significant.

2.6.1. Plumbing Loss

The RF Plumbing losses in the radar are generally considered the losses in microwave components, which are being used in the radar to convey the energy from antenna to the receiver and from transmitter to the antenna. These microwave components are circulator, duplexer, directional coupler, wave-guides and bends etc.

2.6.1.1. Transmission Lines Loss

The losses occurs in transmission line are not one-way loss; it is always two-way loss because these lines are used for reception and transmission both. The theoretically two ways plumbing loss in decibels in radar operating in the S band is approximately 6.8 db.

2.6.1.2. Duplexer Loss

The loss of signal power due to duplexer normally depends upon the type of the duplexer. The duplexer loss is not same as during transmission as during reception of the energy. Manufactures provides the detail of losses occurred in the duplexer.

2.6.1.3. Antenna Losses

There are losses, which are being treated as antenna losses. Some losses of antenna are given below:

1. Antenna beam shape loss,
2. Scanning losses,
3. Radome losses.

2.7. INTEGRATION OF RADAR PULSES

The main purpose of Pulse Integration in Radar is to improve signal to noise ratio. In this operation the signal received from the target is carry on adding till it crosses the threshold level which is just set above expected level of the noise.

Integration is the process of combining *n* samples of a signal, each accompanied by an independent sample of noise or interference, to improve the signal to noise ratio. The samples may be successive pulses in a train or pulses received in parallel channels with independent noise.

A pulse integrator is improvement techniques to address gains in probability of detection by using multiple transmit pulses. This gain will be achieved by inserted in receiving path radar signal processor adding radar returns (thus the word integrator) from different successive pulse periods.

The ability to detect targets is inhibited by the presence of noise and clutter. Both are modeled as random processes; the noise as uncorrelated from sample to sample, the clutter as partially correlated (including possibly uncorrelated) from sample to sample. The target is modeled as either non-fluctuating or a random process that can be completely correlated, partially correlated or uncorrelated from sample to sample (the Swerling models).

In the case of successive signal samples having phase coherence, the integration may be performed by adding voltages at intermediate frequency or at baseband on I- and Q-samples, before envelope detection. This is called *coherent or pre detection integration.* If phase coherence cannot be obtained, or if the complexity and Doppler sensitivity of the coherent integrator is not desired, the signals may be passed through an envelope detector prior to addition of voltages. This is called *video, non-coherent, or post detection integration.*

Envelope detected signals may be converted to digital form and integrated in accumulating registers. If the conversion is to one bit digital form, the process is called *binary integration.* Signal information may also be combined using a simple process of accumulating probabilities of detection over n trials. This may be termed *cumulative integration.*

The signal to noise ratio is improved by integration, greatest improvement being obtained in coherent integration, where the integrated $(S/N)_i = n(S/N)$, and S/N is the signal to noise ratio per sample at the input. For video integration, the improvement is less: $(S/N)_i = n(S/N)/L_i(n)$, where $L_i(n)$ is the integration loss. The binary has an additional loss of about 1.6 dB relative to video integration.

Integration gain is the net improvement in received signal to noise ratio (*SNR*) due to the addition of independent samples of signal and noise available during the radar observation time (time on target).

There are four distinct ways, listed in order of declining integration efficiency and declining complexity of implementation, in which the information from n pulses or samples may be processed to improve the radar detection performance:

1. ***Coherent Integration*:** After coherent demodulation, to the baseband complex-valued (I and Q, or magnitude and phase) data. Combining complex data samples is referred to as coherent integration.

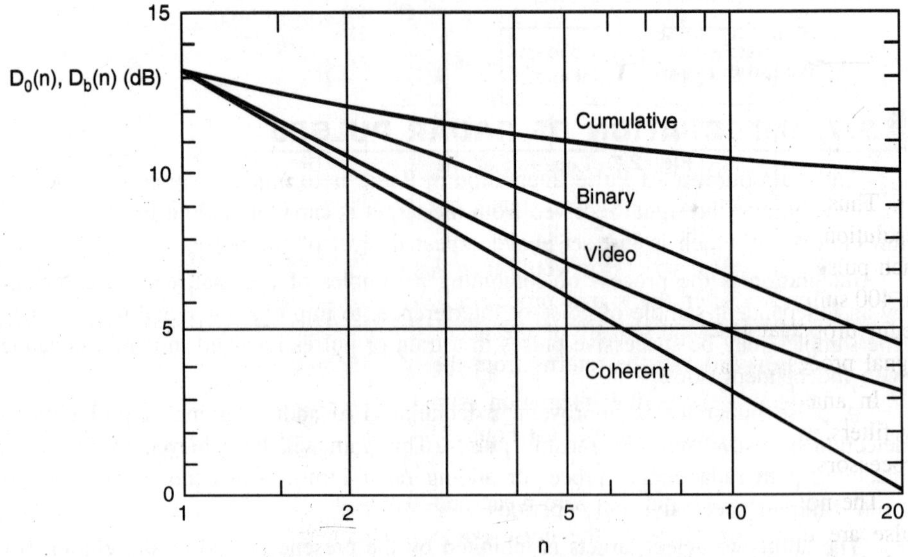

Fig. 2.6. Comparison of Detectability Factors for Four Methods of Integration.

2. ***Non-coherent Integration***: After envelope detection, to the magnitude (or squared or log magnitude) data. Combining magnitude samples after the phase information is discarded is referred to as non-coherent integration.
3. ***Binary Integration***: After threshold detection, to the target present/target absent decisions. This technique is referred as binary integration.
4. ***Cumulative Integration:*** The figure shown below, Fig. 2.6 compares the integration efficiency of each of the four methods in terms of the resultant detectability factor (or SNR) required as a function of n for the conditions noted.

Integration time, also referred to as observation time or time on target, is the time during which the target is illuminated by the radar on each scan, that is

$$t_0 = \frac{\theta_{3az}}{\omega_s} = \frac{t_s \psi_b}{\psi_s}$$

where, θ_{3az} is the 3-dB azimuth beamwidth of the radar, ω_s is the radar scan rate, t_s is the radar frame time, ω_s is the radar's solid angle of search (in steradians), and ψ_b is the solid angle of the radar beam.

For a continuous wave (CW) radar, or a coherent pulse radar, the coherent integration time is not the same as the time on target but is the integration time of the predetection filter, approximately equal to the reciprocal of the filter bandwidth. Subsequent noncoherent integration is possible over the time on target.

1. Coherent Integration

Coherent [predetection] integration occurs when all of the radar pulses n received from a target during the observation time are added in phase before envelope detection. With coherent integration we insert a coherent integrator, or signal processor, between the matched filter and amplitude detector as shown in Fig. 2.7. The signal processor samples the returns from each transmit pulse at spacing equal to the range resolution of the radar set and adds the returns from N pulses. After it accumulates the N pulse sum it performs the amplitude detection and threshold check.

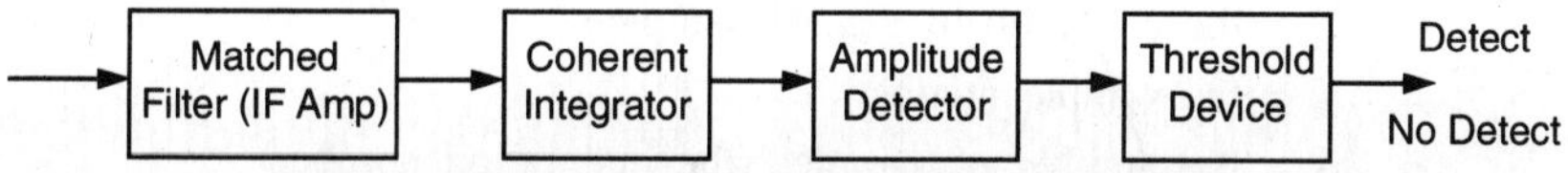

Fig. 2.7. Location of the Coherent Detector.

Thus, for example, if we were interested in a range of 60 km and had a range resolution of 150 m the signal processor would form 60,000/150 or 400 samples for each pulse. The signal processor would then accumulate (add) each of the 400 samples in 400 summers. After the signal processor has summed the first N pulses it would begin dropping older pulses off of the accumulator as new pulses arrive. Thus, the signal processor will add the returns from the most recent N pulses.

In analog processors the integration (summation, accumulation) is accomplished by filters. It is accomplished by Fast Fourier Transformations (FFTs) in digital signal processors.

The noise on each pulse is zero-mean and Gaussian. The noise samples from each pulse are uncorrelated. Thus the noise out of the coherent integrator has the same statistical properties of the noise out of the matched filter from the IF amplifier.

The signal to noise ratio is enhanced by the factor,

$$n = f_r t_0,$$

over that of a single pulse, where f_r is the pulse repetition frequency and t_0 is the integration time. Similarly, for a continuous wave (CW) radar, during the observation time, t_0, there will be

$$n = B_n t_0$$

samples of signals and independent noise added in a coherent integration process, where B_n is the noise bandwidth of the filter. In either case, coherent integration requires that the signal have a predictable phase relationship (coherence) and that the phase response of the filter be such as to bring all of the signal components into the same phase during the integration process. In an ideal coherent integration scheme, the coherent integration gain is exactly n. Coherent integration is sometimes referred to as predetection integration, in that it occurs in the radar receiver before the second (envelope) detector.

We assume that the signal level at the input to the coherent integrator is constant from pulse to pulse. This is indicative of a Swerling Case 0 target Swerling Case 1 target or a Swerling Case 3 target. These signal levels will be added in the integrator. The specific amplitude over the N pulses integrated by the coherent integrator is governed by the probability density function for the specific target type. So we can essentially consider the output of the coherent integrator as the return from a single pulse whose Signal-to-Noise Ratio (SNR) is N times the SNR provided by the radar range equation.

Coherent integration offers no benefit for Swerling Case 2 and Swerling Case 4 targets. This stems from the fact that, that the signal of these targets is not constant from pulse to pulse but, instead, behaves like noise.

Meyer and Meyer treat coherent integration by calculating an equivalent integrated signal-to-noise ratio, is given as

$$\overline{(S/N)_N} = \left(\frac{1}{N}\right)\sum_{k=1}^{N}\left(\frac{S}{N}\right)_k$$

where, $\overline{(S/N)_N}$ = the mean S/N ratio after integration,

$(S/N)_k$ = the S/N ratio of the kth sample

k = the sample number

N = the number of samples (the integration number)

The above relationship gives an average (mean) S/N ratio for the multiple echoes. The integrated S/N is thus N times the mean S/N of the individual hits.

$$(S/N)_N = N.\overline{(S/N)_N}$$

$(S/N)_N$ = the integrated signal to noise ratio

Coherent Integration Losses: There are two losses, first is loss caused by windowing the signal and the second portion of the coherent integration loss involves signal phase.

Most coherent integrations are implemented as either Fourier Transform or correlations. To limit leakage, where signals are in one Doppler or range bin spread to other bins, the signals are windowed prior to the process.

Coherent integration of moving targets requires that the motion be removed before signal summation. This is achieved by subtracting from each consecutive hit a phase equivalent to the target motion between hits. If the motion compensation is perfect, the integrated signal is the linear sum of the components and there is no loss. If not, the summation does not have the maximum possible value.

The total coherent integration loss is given as follows,

$$L_{i(C)} = L_W L_P \qquad \text{(Power Ratio)}$$

$$L_{i(C)}\text{ (dB)} = L_W\text{ (dB)} + L_P\text{ (dB)} \qquad \text{(decibels)}$$

$L_{i(C)}$ = the coherent integration loss

L_W = the signal energy loss caused by the window

L_P = the phase components of the loss

2. Non-Coherent Integrator

Non-coherent integration occurs when n signal samples or pulses are added after passing through an envelope detector which removes any phase relationships. The integrated SNR is increased by a factor approaching n, but decreased by a detector loss due to a loss in information for target detection purposes.

The Non-Coherent Integrator is placed after the amplitude or square law detector as shown in Fig. 2.8. The name non-coherent integration derives from the fact that, since the signal has undergone amplitude or square law detection, the phase information is lost. The non-coherent integrator operates in the same fashion as the coherent integrator in that it sums the returns from N pulses before performing the threshold check.

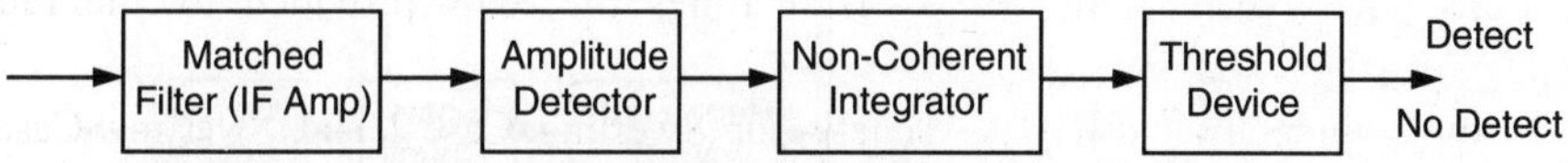

Fig. 2.8. Location of Non-coherent Integrator.

As the signal input to the envelope detector decreases, the small signal suppression loss increases, and further video integration cannot restore the SNR represented by the total signal energy ratio. As a result, there is a requirement for more energy per pulse, which can be considered as integration loss, which is the ratio of total signal energy required of the n-pulse train to that which would have been required if a single pulse had been transmitted and processed, or if a matched filter for the n-pulse train had been used

Cumulative integration refers to the process in which a detection decision with probability P_1 is made on each pulse, resulting in a cumulative detection probability P_c after n pulses :

$$P_c = 1(1 - P_1)^n$$

This process gives far less gain than other video integration techniques, but may represent the only option if the pulses are separated in time by an interval that permits targets to move from one resolution cell to the next, as in scan to scan integration.

3. Binary Integration

Binary integration is a non-coherent integration process in which envelope detected signals are quantized by an initial threshold into one bit binary signals, before being passed to an accumulator. When the accumulator count reaches a second threshold level m, detection is declared. When n pulses are integrated in this way, the optimal second threshold is in the order of $m_{opt} = 1.5\sqrt{n}$ and the integration gain is 1.6 dB lower than that of the n-pulse video integrator. The probability P of detection or false alarm at the output of a binary integrator can be found as a function of the corresponding probability p at its input as

$$P(m/n) = \sum_{j=m}^{n} \frac{n!}{j!(n-j)!} p^j (1-p)^{n-j}$$

The general block diagram of the binary integrator is shown in the Fig. 2.9. If the binary counter is periodically decremented to maintain a low false alarm probability on a continuous stream of input signals, the integrator is known as a moving window (or continuous) binary integrator. Otherwise, it is a batch integrator, in which the counter is set to zero after each group of n pulses.

Although the binary integrator introduces an additional loss relative to the ideal non-coherent integrator, it is much less sensitive to the effects of large, random interference pulses because the energy in a single pulse contributes no more than a single (one) in the binary counter, rather than its large voltage in the linear integrator.

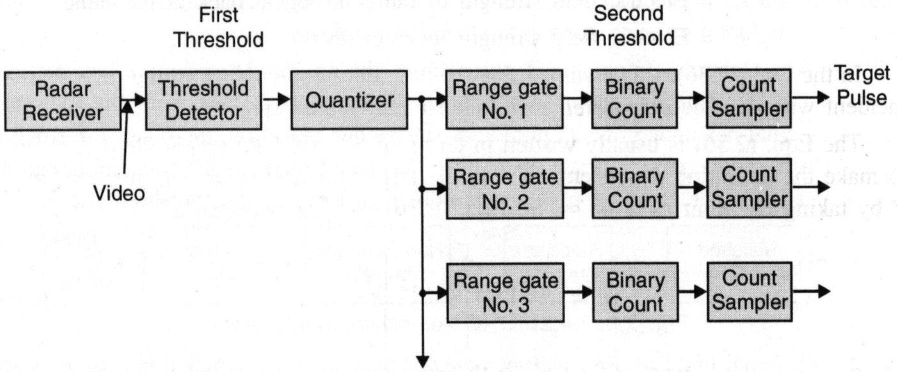

Fig. 2.9. Block Diagram of Binary Integrator.

This type of integrator is also known as the *double threshold detector, m-out-of-n detector, or coincidence detector*, and is widely used in radar signal processors.

The other type of non-coherent detector is implemented as a summer or integrator. In older radars low-pass filters were used to implement them. In newer radars they are implemented in special purpose hardware or the radar computer as digital summers. They operate as described as coherent integrators.

Coherent integration is more effective than non-coherent integration, because more is known about the coherent signal. The integration loss usually ranges from 1.0 to 1.7, and is rarely larger than 2.0, whereas with non-coherent integration, the loss can be as great as the square root of the integration number for very large integrations.

2.8. RADAR CROSS-SECTION OF TARGETS

The radar cross-section σ is the property of a target or scattering object, which is included in the radar range equation to represent the magnitude of echo signal returned to the radar by the target. Radar cross-section RCS is not equal to the physical cross-sectional area of the target. But it is an equivalent area that can be used to relate incident power density at the target to the reflected power density that results at the reciever.

The radar cross-section was defined in the simple radar range equation in terms of following equation

$$\text{Reradiated power density back at the radar} = \frac{P_t G}{4\pi R^2} \cdot \frac{\sigma}{4\pi R^2}$$

In another way, the *RCS* may also be defined as

$$\sigma = \frac{\dfrac{\text{Power reflected toward source}}{\text{Unit solid angle}}}{\dfrac{\text{Incident power density}}{4\pi}} \qquad \text{...(2.35)}$$

$$\sigma = 4\pi R^2 \frac{|E^r|^2}{|E^i|^2} \qquad \text{...(2.36)}$$

where R = Range to the target

E^r = Electric field strength of the echo signal back at the radar

E^i = Electric field strength incident on the target.

In the Eqn. (2.36), it is assumed that the target is far enough from the radar that the incident wave can be considered to be planar rather than spherical.

The Eqn. (2.36) is usually written in terms of electric field amplitude and in order to make the definition dependent only on the target characteristics, we eliminate range R by taking the limit as $R \rightarrow \infty$. So Eqn. (2.36) may be rewritten as

$$\sigma = 4\pi \lim_{R\to\infty} \left[4\pi R^2 \frac{|E^r|^2}{|E^i|^2} \right] \qquad \text{...(2.37)}$$

Sometimes the *RCS*σ is said to be a fictional area that intercepts a part of the power incident at the target which, if scattered uniformly in all direction, produces an echo power at the radar equal to that produced at the radar by the real target. But the real targets do not scatter the incident energy uniformly in all directions.

The power scattered from a target in the direction of the radar receiver *i.e. RCS*, can be calculated by solving Maxwell's equations using the proper boundary conditions.

The *RCS* is just defined as a single real scalar number. The above definition of *RCS* uses a single polarization for the transmitted and receiving wave usually the same. The polarization state of a transverse EM plane wave is a two dimensional vector and two orthogonal polarization basis vector are required to fully describe the wave. The commonly there are two types of polarization basis one is linear (horizontal and vertical polarization) and circular (right and left circular polarization). To understand the effect of polarization, *RCS* must be generalized to the polarization scattering matrix (*PSM*) *S* that relates the complex amplitudes of the incident and backscattered fields.

The following relation for a radar using a linear polarization basis is given by Knott, Shaeffer and Tully, 1985; Mott, 1986; Holm, 1987.

$$\begin{bmatrix} E^r_{(H)} \\ E^r_{(V)} \end{bmatrix} = \begin{bmatrix} S_{HH} & S_{HV} \\ S_{VH} & S_{VV} \end{bmatrix} \begin{bmatrix} E^i_H \\ E^i_V \end{bmatrix}$$

$$= S \begin{bmatrix} E^r_H \\ E^r_V \end{bmatrix} \qquad \text{...(2.38)}$$

The target back scattering characteristics are described by four complex number rather than of a single real number. If we assume that both radar transmitter and receiver have only the vertical components, then the *RCS* may be expressed as follows :

$$\sigma = |S_{VV}|^2 \qquad \text{...(2.39)}$$

In general radars area designed to measure the full Polarization Scattering Matrix (PSM). There are other designs measure the magnitude of the elements of the PSM, but not the phases or the magnitudes of two of the PSM elements, which are known as polarimetric measurements. The polarimetric measurements can be used for the analysis of various target.

2.9. STATISTICAL DESCRIPTION OF RCS

For the real targets, the RCS cannot be effectively modelled as a simple constant but in general it is a complex function of aspect angle, polarization and frequency. At normal incidence the radar cross-section of a flat plate of area A in the optics region is given by

$$\sigma = \frac{4\pi A^2}{\lambda^2} \qquad \text{...(2.40)}$$

which can give quite large values for a modest size plate. For example, if wavelength is in order of 3 cm or frequency in X-band, a square plate 0.3 m on a side has a RCS of 113 m^2 when viewed normal to the surface. The magnitude of the backscatter from a flat plate reduces rapidly alongwith the angle from the normal.

The radar cross-section of a dihedral or trihedral corner reflector (a dihedral corner reflector is a structure formed by two flat faces perpendicular to each other which returns the incident wave back to its source and a trihedral corner reflector is made up of three faces mutually perpendicular to each other) of projected area also given by the same expression as the flat plate but it applies over a much larger viewing angle. A conducting trihedral corner reflector as shown in Fig. 2.10 is very often used as a calibration target in field measurements. The RCS of a trihedral corner reflector when viewed along its axis of symmetry (seeing into the corner) may be determined by

$$\sigma = \frac{12\pi A^4}{\lambda^2} \qquad \text{...(2.41)}$$

The RCS is increasing with increasing frequency.

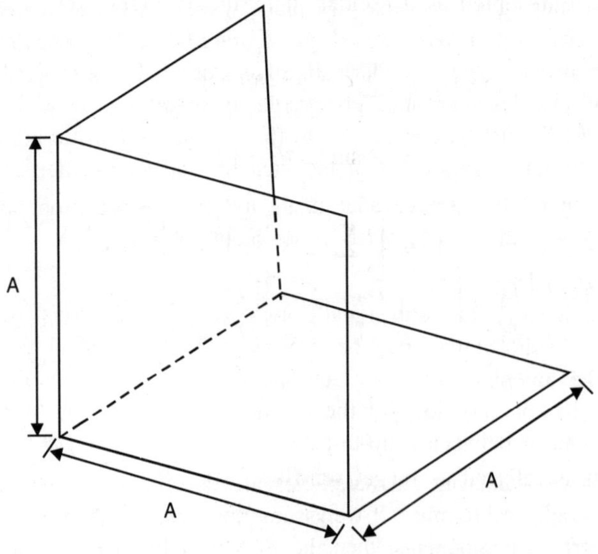

Fig. 2.10. A Trihedral Corner Reflector.

A statistical description of radar cross-section used to study the complicated behaviour of complex targets. It means that the radar cross-sections of the scatters within a single resolution cell is considered to be a random variable with a specified probability density function. The RCS of complex targets such as aircraft, ships, missiles, ground vehicles, buildings and terrian can vary considerably depending on the aspect angle and frequency. The range equation is used to estimate the mean RCS and one of a variety of probability density function are used to describe the statistical behaviour of the RCS.

Common Statistical Models for RCS

(*i*) Rayleigh/exponential, chi-square of degree 2.
(*ii*) Non-fluctuating, Marcum, Swerling 0, or Swerling 5
(*iii*) Chi-square of degree 4
(*iv*) Chi-square of degree 2 m, weinstock
(*v*) Rice or Rician, non-central Chi-square of degree 2
(*vi*) Weibull
(*vii*) Log-normal

2.10. RCS FLUCTUATIONS OF TARGETS

Generally all radar targets objects produce echo signals that vary in amplitudes either in power or cross-sectional terms. A small change in viewing aspect of a radar target such as an aircraft or ship can result in major change in the radar cross-section. Complex targets are made up of a number of individual scattering centers or scatterers. The scatterers for an aircraft might be the engines, cockpit, nose, wings, tails and external stores. The echo from each scattering center has an amplitude and phase that is independent of the amplitude and phase of the echoes from other scattering centers. The phase of each scatteres is determined primarily by the distance of the individual scattering center from the radar. Finally at the radar all the echoes from the individual scatters add vectorially to form a resultant amplitude and phase. The echo signal $S_r(t)$ may be written as following if the target can be represented as a collection of independent point scatters,

$$S_r(t) = \sum_{i=1}^{N} a_i \sin(2\pi ft + \phi_i) \qquad ...(2.42)$$

$$= A \sin(2\pi ft + \phi)$$

where

$$A = \left[\left(\sum_i a_i \sin\phi_i\right)^2 + \left(\sum_i a_i \cos\phi_i\right)^2\right]^{1/2}$$

In practice, however the echo signal from a target in motion is almost never constant. Variation in the echo signal may be caused by meteorological conditions, the lobe structure of the antenna pattern or equipment instabilities. But the main source of fluctuation is that of variations in the target cross-section. The amplitudes can vary from scan-to-scan or between echo-to-echo due to aspect changes relative to the radar. This variations is referred as **target scintillation.**

2.11. TARGET FLUCTUATION MODELS

A simple target model is shown in Fig. 2.10, which is representing a reflective structure of a satellite or an aircraft. This figure is to that given in Van Trees (1971). If the direction of signal propagation is along the *x*-axis and the target positions are shown in Fig. 2.11(*a*, *b*, *c*). It is assumed that the target in Fig. 2.11, is illuminated with a long pulse of duration *T* whose envelope is given in Fig. 2.11(*a*) and the received signal envelope is shown in Fig. 2.11(*b*).

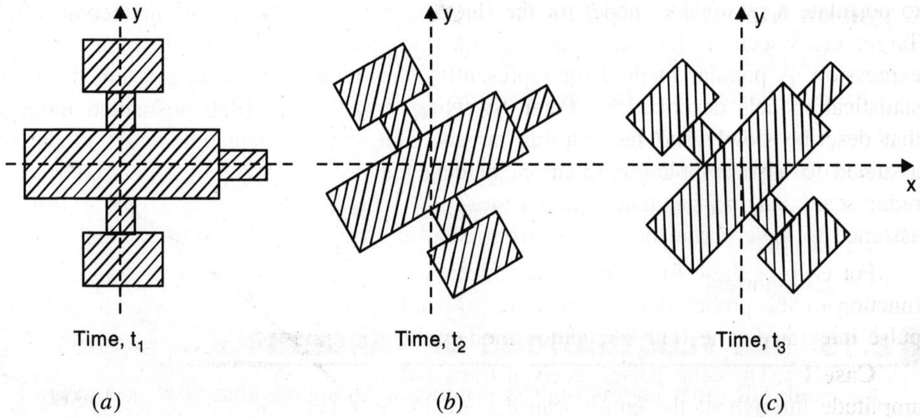

Fig. 2.11. Representatives of Target Orientations at Different Times.

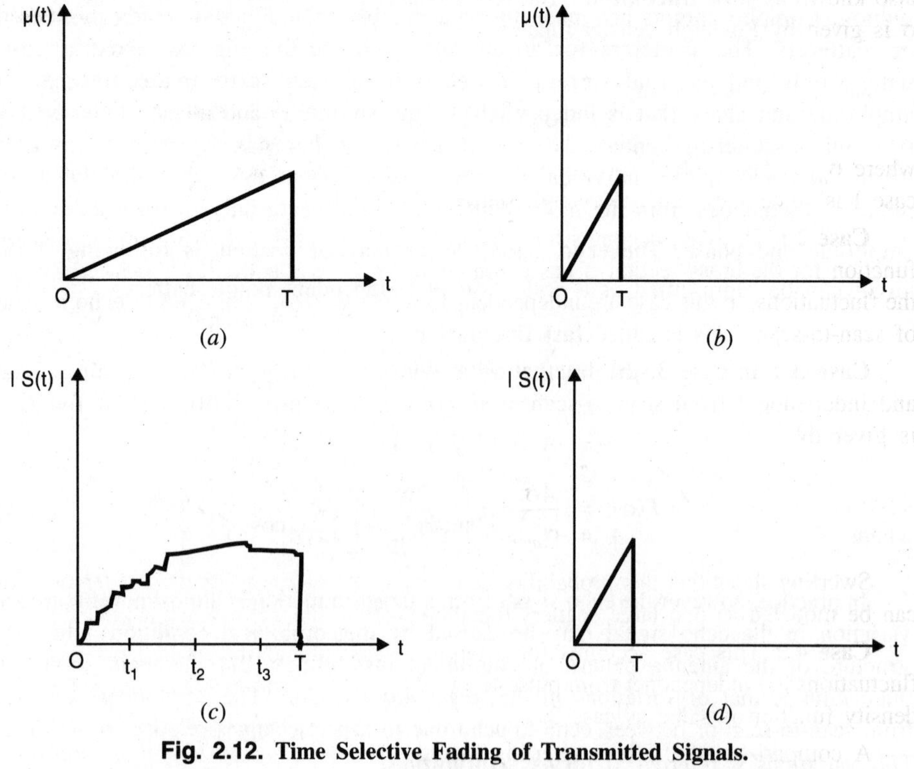

Fig. 2.12. Time Selective Fading of Transmitted Signals.

If the received envelope is distorted as shown in Fig. 2.12(*b*), the target is fluctuating, with time as the target changes its orientation. This varying time attenuation of the received envelope is often called **time-selective fading.** On the other hand, if a short pulse is transmitted as in Fig. 2.12(*c*) and an undistorted signal envelope is received as in Fig. 2.12(*d*), the target can be considered to be slowly fluctuating (Van Trees 1971).

2.12. SWERLING MODELS

Most radar situations are of two complex a nature to warrant obtaining complete data. A more economical method to assess the effects of a fluctuating cross-section is to postulate a reasonable model for the fluctuations and to analyze its mathematically. Target cross-section fluctuation are complex to quantify by a simple mathematical expression. A popular method for representing the fluctuations of targets are the four statistical models described by **Peter Swerling.** Swerling in 1960 postualted models that describe slowly and fast varying targets. The slowly fluctuating target model is assumed to have complete correlation, or dependence, between echo signals during a radar scan, but independent with scan-to-scan. The fast fluctuating target model is assumed to have partial correlation from echo-to-echo instead of scan-to-scan.

For each of these four models, he calculated the signal to noise ratio required as a function of the probability of detection, probability of false alarm and the number of pulse integrated. The four fluctuation models are as follows :

Case 1 : The echo pulses received from a target on any one scan area of constant amplitude throughout the entire scan but are independent (uncorrelated) from scan to scan. A target echo fluctuation of this type is known as **scan-to-scan fluctuation.** It is also known as **slow fluctuation.** The probability density function for the cross-section σ is given by Rayleigh density function

$$P(\sigma) = \frac{1}{\sigma_{av}} \exp\left(-\frac{\sigma}{\sigma_{av}}\right) \sigma \geq 0 \qquad \text{...(2.43)}$$

where σ_{av} = The average cross-section overall target fluctuation. The target model in case I is sometimes called **Rayleigh scatterer.**

Case 2 : This case accounts for a target to fast fluctuation. The probability density function for the cross-section σ has a similar distribution function as that in case-I but the fluctuations, in this case are independent from pulse-to-pulse or echo-to-echo instead of scan-to-scan. This is called **fast fluctuation.**

Case 3 : In case 3, the fluctuation is considered to be constant within a scan and independent from scan-to-scan as in conj, but the probability density function is given by

$$P(\sigma) = \frac{4\sigma}{\sigma_{av}^2} \exp\left(-\frac{2\sigma}{\sigma_{av}}\right) \sigma \geq 0 \qquad \text{...(2.44)}$$

Swerling states that this probability density function is representative of targets that can be modeled as one large scatters together with a number of small scatters.

Case 4 : This case accounts for a target of fast fluctuation as in LAN-2 where fluctuations are independnet from pulse-to-pulse instead of scan-to-scan. The probability density function is same as case 3.

A comparison of the four swerling models of the fluctuated target and the non-fluctuating case, here called as 5 is illustrated in Fig. 2.12. In Fig. 2.12, we have $n = 10$ bits integrated post detection and false alarm number of 10^8 (inverse of the false alarm probability). When the detection probability is large, all four cases in which the target cross-section is not constant require greater signal to noise ratio than the constant cross-section of case 5.

The probability density function given by Eqn. (2.42) and (2.44) are special cases of the **chi-square, or gamma, distribution** with $2n$ degrees of freedom ($2n$ DoF). **Chi-square distribution** is a general approximation of target models. Specifically

$$P(\sigma) = \frac{n}{\sigma_{av}(n-1)!}\left(\frac{n\sigma}{\sigma_{av}}\right)^{n-1} \exp\left(\frac{-n\sigma}{\sigma_{av}}\right), \sigma \geq 0 \qquad ...(2.45)$$

The envelope of σ is taken as a Rayleigh random variable whose average is σ_{av}. When the unresolved target echo results from many scatterers of comparable size adding vectorically with random phases, then the echo amplitude is Rayleigh distributed (and the echo energy is thus exponentially distributed) (Heering 1977). Unlike in statistical texts where index n is only allowed to be an integer, n in this instance must be positive and can also be a real number when applying to target cross-section models.

Nakagemi in 1960 gave a more generalized chi-square model as

$$P(\sigma) = \frac{2n^n \sigma^{2n-1}}{\sigma_{av}^n \Gamma(n)} \exp\left(\frac{-n\sigma^2}{\sigma_{av}}\right), \sigma > 0 \qquad ...(2.46)$$

When $\Gamma(.)$ is the gamma function of (.) and n can be real or integer 1.

There is another model is the **Rician Model (Rice 1944).** The Rician model is suitable for the case of one dominant signal in the presence of many other small signals. It is given as

$$P(x, \hat{x}) = \begin{cases} \dfrac{1}{\psi_0} I_0\left[2\sqrt{\dfrac{S_x}{\psi_0}}\right] e^{-S-\frac{x}{\psi_0}}, & x > 0 \\ 0 & x \leq 0 \end{cases} \qquad ...(2.47)$$

where S = Ratio of steady reflector's RCS to be combined average cross-section of Rayleigh scatterers

$\hat{x} = \psi_0(1+S)$

$\sigma = \psi_0\sqrt{1+2s}$

ψ_0 = Mean value of Rayleigh component of x

I_0 = Modified Bessel functions of the first kind of zero order.

2.13. ANTENNA PARAMETERS

The important parameters of radar antennas, discussed here in brief. Which are as follows :

1. Directivity or Directive gain.
2. Power gain.
3. Effective receiving aperture or total scattering cross-section.
4. Polarizations.
5. Side lobes.

2.13.1. Directivity or Directive Gain

The directivity of an antenna is given by the ratio of the maximum radiations intensity (power per unit solid angle) $u(\theta, \emptyset)$ mass to the average radiations intensity *U*av (average over a sphere) or it may be defined as at a certain distance from the antenna the directivity may be expressed as the ratio of the maximum to the average polyting vector.

$$G_D = \frac{\text{Maximum radiation}}{\text{Average radiation}} \qquad ...(2.48)$$

$$G_D = \frac{u(\theta, \emptyset)_{\text{Max}}}{U\text{av}} \text{ or } \frac{S(\theta, \emptyset)_{\text{Max}}}{S\text{av}} \qquad ...(2.49)$$

The smaller the beam solid angle, the greater the directivity.

The ability of radar antenna to concentrate energy in a particular direction is known as **gain**.

There are two related terms of antenna gain are the power gain and the directive gain. Both the terms are of interest to the radar system engineer. The directive gain or directivity is related to radiation pattern of the antennas. Whereas the power gain is more suitable related to the radar range equations.

The power radiated from an antenna per unit solid angle is called the radiation intensity u (watts per steradian or per square degree). A radiation intensity pattern may be plot between radiation intensity and the angular co-ordinate, this plot is known as a **radiation-intensity pattern**.

The power density, which may be desired as power per unit area, can be plotted as a function of angle is known as a power pattern. The terms, power density pattern and radiation intensity pattern are identical when plotted on a relative basis that is when the maximum is normalized to up to a value of unity. When both are plotted on a relative basis is called the antenna **radiation pattern**.

The directive gain may be written as the average radiation intensity over a solid angle of 4π radians is equal to the total power radiated divided by 4π

$$G_D = \frac{(\text{Maximum power radiated / unit solid angle}) \times 4\pi}{\text{Total power radiated}} \qquad ...(2.50)$$

With the help of above equation. We can find out the directive gain from the radiation pattern.

The directive gain equation can be written as,

$$G_D = \frac{4\pi}{B} \qquad ...(2.51)$$

In above equation 4π is denoted the maximum power per unit solid angle approximately and the total power radiated is find by integrating the volume contained under the radiation pattern which is also known as beam area (B).

The angular range of the antenna pattern in which the half of the maximum power at least is still emitted is described as a Beam Width. Bordering points of this may or lobe are therefore the points at which the field strength has fallen in the room around 3 dB regarding the maximum field strength. This angle is then described as an beam width or half width of the antenna.

The beam area (B) is the solid angle from when adds the total power radiated would pass if the power per unit solid angle were equal to $P(\theta, \emptyset)$ max over the beam area. This solid angle can often be described approximately in term of angles subtended by the half power points of the main to be in the two principles planes as given by

$$B \simeq \theta B \times \varnothing B. \qquad ...(2.52)$$

Now by putting the value of B in the above equation, we get

$$GD\ \varnothing = \frac{4\pi}{(\theta B \varnothing B)} \qquad ...(2.53)$$

Where θB and $\varnothing B$ are the half power beam width into principal planes such as horizontal and vertical respectively, minor lobes are neglected. If the half power beams widths are measured in radians.

If the half power beam width are measured in degrees, and minor lobes effects are neglected then this can be written as

$$G_D = \frac{4\pi}{\theta B \varnothing B} \approx \frac{41000}{(\theta B^\circ \times \varnothing B^\circ)} \qquad ...(2.54)$$

($4\pi sr$ = 4125 square degree)

Suppose as antenna has a main lobe with both half-power beam widths (HPBWs = 2o°) its directivity gain is approximately

$$G_D = 4\pi sr/B(sr) \sim 41000(\text{deg}^2)/(\theta B^\circ \varnothing B^\circ)$$
$$= 41000(\text{deg}^2)/(2o^\circ \times 2o^\circ)$$
$$G_D \simeq 103$$

Or $G_D \simeq 20$ dB (dB above isotrope)

It means that the antennas radiates a power in the direction of the main lobe maximum which is about 100 times as would be radiated by a non-directional (isotropic) antenna for the same power point.

2.13.2. Power Gain

The power gain, which is denoted by (G), includes the effect of the antenna losses and any other related losses, which reduces the antenna efficiency. In the directive gain definition we have not considered the losses such as ohmic (resistance) heating, RF heating or mismatched antenna, we have considered only the radiation pattern. But these losses are taken into account while calculating the power gain. It may be defined as the ratio of maximum radiation intensity from particular antenna to the radiation intensity from (loss less) isotropic source with some power input.

This defination must be taken care while using radar range equation since its includes the losses introduced by the antenna. The directive gain is important for resolution consideration, average and the accuracy, which is related to the antenna beam width.

The directive gain is always greater than power gain. If there is no losses, the both the gain (power and Directive) are more or less equal to each other. Both are related by the radiation efficiency factor (ρ_r) and can be defined as follow:

$$G = \rho_r \times G.D. \qquad ...(2.55)$$

The theory of antenna related to the directivity and power gain is same for the transmission and receiver antenna in radar.

2.13.3. Effective Aperture or Total Scattering Cross Section

One of the important parameter related to the gain is the effective receiving aperture or total scattering cross-section or effective area. It may be considered as a measure of the effective area presented by the antenna to the incident wave.

We know from the radar range equation that

$$G = 4\pi A_e/\lambda^2 \quad ...(2.56)$$

Where λ = Wave length

A_e = Effective area or effective aperture area

ρ_a = Antenna aperture efficiency

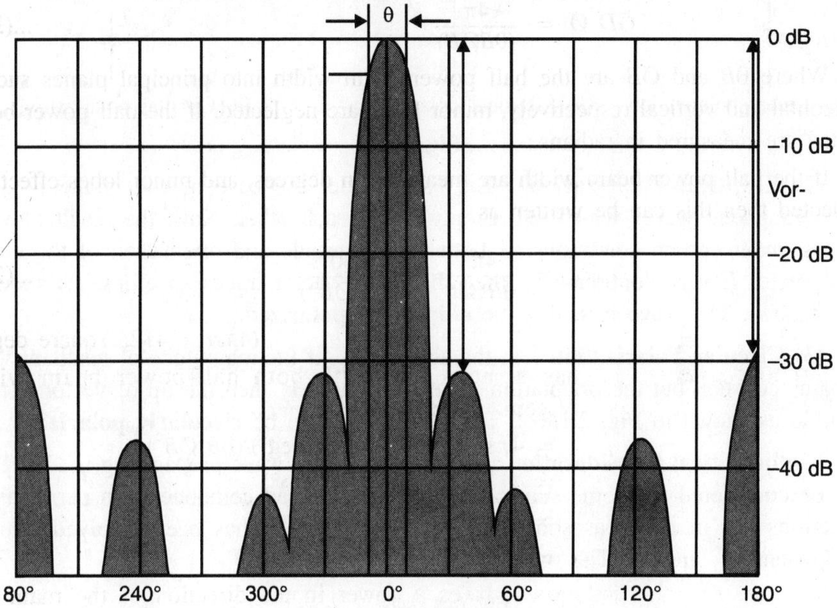

Fig. 2.13. Antenna Pattern in a Rectangular-Coordinate Graph.

The effective area (A_e) may be defined as the product of the antenna aperture effiency (ρ_a) and the physical area of antenna.

$$A_e = \rho_a \times A \quad ...(2.57)$$

$$G = 4\pi \times (\rho_a \times A)/\lambda^2 \quad ...(2.58)$$

2.13.4. Wave Polarization

We have considered the case of a plane electromagnetic wave say propagating in Z-direction. Then the electric vector E is in X-direction while magnetic vector H is in K-direction. There is an orthogonal (perpendicular) relationship between E, H and Poynting vector P. To describe an electromagnetic wave completely, we should not only include the parameters such as its wavelength, phase velocity and power but also the *instantaneous orientation of its field vectors.* It is important to mention here that if the magnitude and direction of E are known then the magnitude and direction of H are automatically specified. Thus, it is sufficient to specify the quantities related to electric vector E.

*The **wave polarization** is defined as the orientation of electric field vector E at a given instant of time in space.*

There are three types of polarization :

(1) Linear Polarization : A plane electromagnetic wave in which electric vector vibrates simple harmonically along a fixed straight line perpendicular to the direction of propagation without changing its orientation is called *plane polarized* electromagnetic wave as shown in Fig. 2.14(a). The wave is travelling out of page in Z-direction.

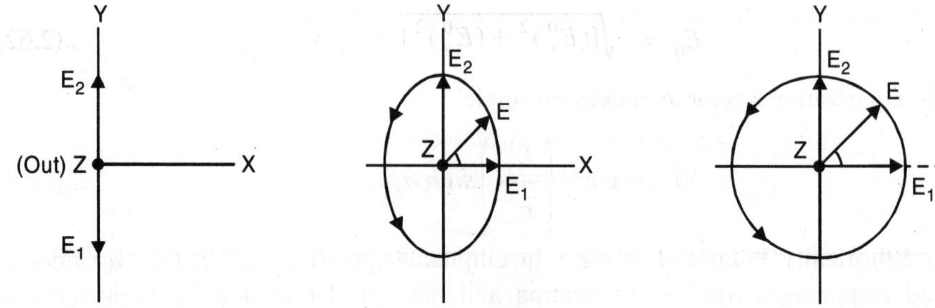

(a) Linear polarization (b) Elliptical polarization (c) Circular polarization.

Fig. 2.14.

(2) Elliptical Polarization : Let us consider the case of two plane polarized electromagnetic waves which superimpose over each other. Now, the resultant vector rotates under certain conditions. If both-the magnitude and orientation of the electric field vector E vary continuously, then the tip of vector traces an ellipse as shown in Fig. 2.14(*b*). The wave is said to be **elliptically polarized.**

(3) Circular Polarization : In the above case, if the magnitude of resultant vector remains constant but the orientation varies continuously then the tip of vector *E* traces a circle as shown in Fig. 2.14(*c*). The wave is said to be **circularly polarized.**

Mathematical Consideration : A plane polarized wave propagating in *Z*-direction can be considered to be the sum of two plane polarized components in perpendicular directions but in phase as shown in Fig. 2.15. Here *E* has been resolved into two components E_1 and E_2. The two components are given by

$$E_x = E_x^0 e^{j(\omega t - kx)}$$

or
$$E_x = \hat{a}_x E_x^0 e^{j(\omega t - kx)} \quad ...(2.59)$$

Similarly,
$$E_y = \hat{a}_y E_y^0 e^{j(\omega t - kx)} \quad ...(2.60)$$

Here, $\hat{a}_x$ and $\hat{a}_y$ are unit vectors along *X*- and *Y*-axis respectively.

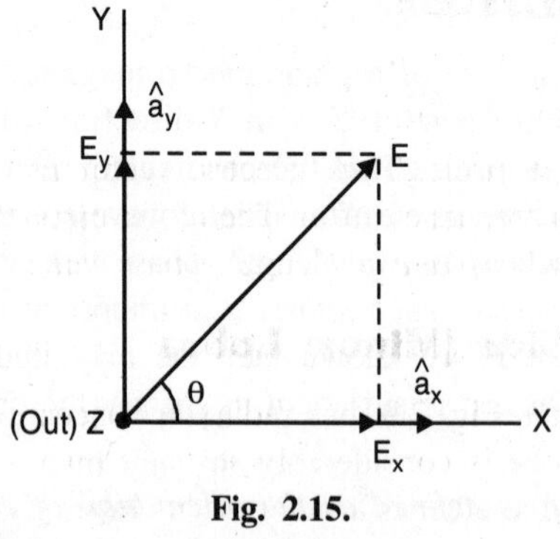

Fig. 2.15.

(1) Plane Polarized Wave : The general solution of plane polarized wave is given by

$$E(r, t) = (\hat{a}_x E_x^0 + \hat{a}_y E_y^0) e^{j(\omega t - kx)} \quad ...(2.61)$$

Here, E_x^0 and E_y^0 must have the same phase if $E(r, t)$ is to be plane polarized. The magnitude E_0 is given by

$$E_0 = \sqrt{[(E_x^0)^2 + (E_y^0)^2]} \quad ...(2.62)$$

The polarization vector E makes an angle

$$\theta = \tan^{-1}\left(\frac{E_y^0}{E_x^0}\right) \text{ with } \hat{a}_x \quad ...(2.63)$$

(2) Elliptically Polarized Wave : In elliptically polarized wave, the two linearly polarized components (one in X-direction and other in Y-direction) are not equal in magnitude. As the magnitude is real, we can expression them as

$$E_x(r, t) = E_1 \cos(\omega t - kz) \quad ...(2.64)$$

and

$$E_y(r, t) = E_2 \sin(\omega t - kz) \quad ...(2.65)$$

where E_1 is amplitude of linearly polarized light in X-direction and E_2 is the amplitude of wave linearly polarized in Y-direction.

The resultant of Eqns. (2.64) and (2.65) is given by

$$\frac{E_x^2}{E_1^2} + \frac{E_y^2}{E_2^2} = 1$$

This is equation of ellipse. Therefore, elliptically polarized wave is the resultant of two plane perpendicular polarized waves of unequal amplitudes.

(3) Circularly Polarized Wave : The components of actual electric field obtained by taking real part of Eqn. (2.60) are given as

$$E_x(r, t) = E_0 \cos(\omega t - kz) \quad ...(2.66)$$

and

$$E_y(r, t) = E_0 \sin(\omega t - kz) \quad ...(2.67)$$

In case of circularly polarized wave, the amplitude of two components is the same, *i.e.*,

$$E_1 = E_2 = E_0$$

The resultant of Eqns. (2.66) and (2.67) is given by

$$\frac{E_x^2}{E_0^2} + \frac{E_y^2}{E_0^2} = 1$$

or

$$E_x^2 + E_y^2 = E_0^2 \quad ...(2.68)$$

Eqn. (2.68) represents a circle. Thus, electric vector E, at a fixed point in space, sweeps around a circle at a frequency of ω. Therefore, *circularly polarized wave is the resultant of two perpendicular plane polarized waves with equal amplitude.*

2.13.5. Major and Side (Minor) Lobes

The pattern shown in the Fig. 2.9, has radiation concentrated in several lobes. The radiation intensity in one lobe is considerably stronger than in the other. The strongest lobe is called a major lobe; the other is a side (minor) lobes. Since the complex radiation patterns associated with arrays frequently contain several lobes of varying intensity. In general, major lobes are those in which the greatest amount of radiation occurs. Side or minor lobes are those in which the radiation intensity is least.

In general side lobes are not at all desirable but in radar up to some extent side lobes are permitted. Low side lobes are generally required for radar system but if it is too large than the maximum portion of the radiated energy contained in the side lobes, which result, reduction in main beam energy and in turn reduces the gain also. If the side lobes are high than there are chances of getting false target indication because strong echo signal may enter through side lobes and mislead the radar operator. It is very easy to jam the radar, if its side lobes too high and it is more likely to get interference from nearby other transmitters.

The level of side lobes should of the order of 20 to 30 dB below the main beam and it can be achieved easily with the practical antenna. To achieve very low side lobes is quite difficult practically due to limitation imposed by the nature but theoretically there is no reason why they should not be possible.

2.13.6. Front-to-Back Ratio

The front-to-back ratio of an antenna is the proportion of energy radiated in the principal direction of radiation to the energy radiated in the opposite direction. A high front-to-back ratio is desirable because this means that a minimum amount of energy is radiated in the undesired direction.

2.13.7. Beam Efficiency

Beam efficiency is also an important parameter to judge the quality of transmitting and receiving antenna. In terms of power, the **beam efficiency** can be defined as the ratio of power transmitted (may be received) within cone angle 61 to the power transmitted (or received) by the antenna and it is dimensionless.

$$\text{Beam efficiency} = \frac{\text{Power transmitted (received) within cone angle } \theta_1}{\text{Power transmitted (received) by the antenna}}$$

where the major lobe is directed along the z-axis where $\theta = 0$ and θ_1 is the half angle *of the* CMC *within which* the percentage of the total power is to be found, as shown in Fig. 2.16.

The beam efficiency should be high usually 90. It is necessary for the antenna used in radar, astronomy and other applications where reception through the side lobes is not desired.

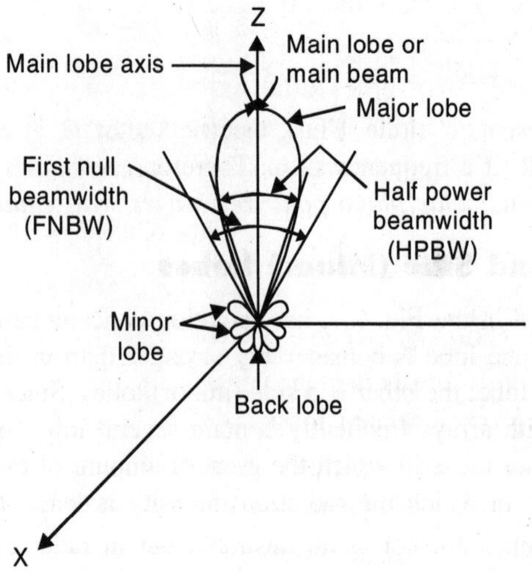

Fig. 2.16. Radiation Lobes and Beamwidth of Antenna Pattern.

In other terms, the **beam efficiency** can be defined as the ratio of the main beam area to the total beam area. It is denoted by ε_M,

$$\varepsilon_M = \frac{\Omega_M}{\Omega_A} = \text{Beam efficiency} \quad ...(2.69)$$

where Ω_A is the total beam area or beam solid angle and consists of the main beam (Ω_M) and the main lobe area (Ω_m).

SUMMARY

- The main purpose of Pulse Integration in Radar is to improve signal to noise ratio.
- Integration is the process of combining *n* samples of a signal, each accompanied by an independent sample of noise or interference, to improve the signal to noise ratio. The samples may be successive pulses in a train or pulses received in parallel channels with independent noise.
- *Integration gain* is the net improvement in received signal to noise ratio (*SNR*) due to the addition of independent samples of signal and noise available during the radar observation time (time on target).
- Coherent [pre-detection] integration occurs when all of the radar pulses n received from a target during the observation time are added in phase before envelope detection.
- Non-coherent integration occurs when *n* signal samples or pulses are added after passing through an envelope detector which removes any phase relationships.
- Binary integration is a non-coherent integration process in which envelope detected signals are quantized by an initial threshold into one bit binary signals, before being passed to an accumulator.
- The radar cross-section θ is the property of a target or scattering object, which is included in the radar range equation to represent the magnitude of echo signal returned to the radar by the target.
- A popular method for representing the fluctuations of targets are the four statistical models described by **Peter Swerling**.
- The directivity of an antenna is given by the ratio of the maximum radiations intensity (power per unit solid angle) $u(\theta, \text{Ø})$ mass to the average radiations intensity U_{av} (average over a sphere).
- The power density, which may be desired as power per unit area, can be plotted as a function of angle is known as a power pattern.
- The power gain, which is denoted by (*G*), includes the effect of the antenna losses and any other related losses, which reduces the antenna efficiency.
- The wave polarization is defined as the orientation of electric field vector *E* at a given instant of time in space.
- Beam efficiency is also an important parameter to judge the quality of transmitting and receiving antenna.

REVIEW QUESTIONS

1. Differentiate between range resolution and cross range resolution.
2. What do you understand by integration of radar pulses? How these effect the radar performance?
3. Discuss the different type of integrators in brief.
4. What is coherent integration and how it is differ to non-coherent integration?
5. Discuss radar cross-section of a target and how it is related to the radar range.
6. Discuss radar cross-section fluctuations in brief.
7. Discuss antenna parameters.

❑❑❑

CHAPTER

3

MTI AND PULSE DOPPLER RADAR

3.1. INTRODUCTION

INSIDE THIS CHAPTER

The radars discussed till now were required to detect targets in the presence of noise. But in practical radars have to deal with more than receiver noise when detecting targets while they can also receives echoes from the natural environment such as land, sea and weather. These echoes are called **clutter**, since they tend to clutter the radar display with unwanted information's. Clutter echoes signal has greater magnitude then echo signal receives from the aircraft. When an aircraft echo and a clutter echo appear in the same radar resolution cell, the aircraft might not be detected. But the Doppler effect permits the radar to distinguish moving targets in the presence of fixed target even the echo signal from fixed targets has comparatively greater magnitude than the moving targets such as aircraft.

3.2. MOVING TARGET INDICATOR RADAR

Echo signals from fixed targets are not shifted in frequency, but the echo from a target moving with relative velocity V_r will be shifted in frequency by an given by the Doppler formula, $f_d = 2v_r/\lambda$, when λ is wave length of the transmitted signal.

Pulse radar, which makes use of the Doppler shift for detecting moving targets is either, MTI radar or a pulse Doppler radar. The MTI radar has a PRF low enough to not have any range ambiguities $R_{un} = c/f_p$ (such as multiples around echo). It does however have many ambiguities in the Doppler domain (such as blind speed).

The pulses Doppler radar on the other hand have a PRF large enough to avoid doppler ambiguities but it can have numerous range ambiguities. These are also a medium PRF pulse doppler that accepts both range and doppler ambiguities.

There is a type of MTI which does use the informations directly from the doppler but it observing the area map scan to scan. Only moving objects which are changes their position scan to scan from mapping operations are displayed. This type of MTI is known as **Area MTI**.

Figure 3.1 shows the five successive return on MTI radar. This sweep shown several fixed targets and a moving target. On the basis of single sweep. Moving targets cannot be distinguished from fixed targets. Echoes from fixed targets remain constant throughout, but echoes from moving targets varies sweep to sweep at a rate corresponding to the doppler frequency.

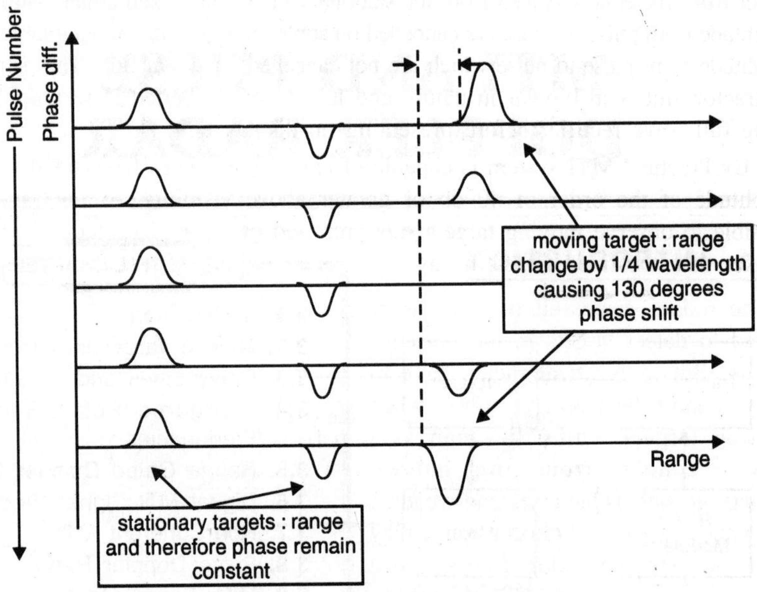

Fig. 3.1. Five Sequential Returns from Pulse Comparison Output.

3.2.1. A Simple Block Diagram of MTI Radar

A schematic diagram of MTI radar is shown in Fig. 3.2. This diagram has got two channels, one is receiver channel and other is coherent oscillator (COHO Channel). A receiver channel is identical to the standard radar receiver except the local oscillator which is designed to have good frequency stability and a limiter is included in the IF stage so that the amplitude of echo signal fed to the phase detector should have the same amplitude. The reference signal is provided by the COHO oscillator (Coherent Oscillator) and is a stable oscillator (STALO) operating at a frequency that is near to the difference between transmitter and Stalo frequencies.

The phase of each transmitter pulse is randomly related to the phase of the last pulses in the radar using magnetrons as transmitter. It is essential that the phase of the echo oscillator be matched with the phase of the transmitter pulse is injecting in to Coho channel. If the frequency of the Coho oscillator is stable, its phase in the time interval between pulses is then related to the phase of the oscillations of the last outgoing pulse in exactly the same way for successive pulses. The phase detector used in MTI radar compare the phases of the Coho oscillator and the amplitude limited intermediate frequency echo signal. The phase detector output for successive pulses may be subtracted from each other by employing a single delay line. A delay line provides the delay of one PRP, it may employ a mercury delay line or quartz transmission line, upon which are impressed mechanical vibrations lying in the frequency range 10 to 20 MHz, modulated by the video signals. The line length is such as to introduce a delay corresponding exactly to the period between successive pulses. It may be ensured by using the delayed outgoing pulse to trigger off the next outgoing pulse. The delayed and undelayed echoes are subtracted by first reversing the polarity of one of them and adding.

The phase comparison is carried out at IF stage instead of doing at the much higher transmitted frequency. The output of the phase detector is fed as the input to the delay line canceller, which is acting as a high pass filter. It will separate the moving target from the echo received from the stationary target. The fixed targets with the constant amplitude from pulse-to-pulse are cancelled on subtraction. But the moving target has varying amplitude from pulse-to-pulse which are not cancelled on subtraction. The output from the subtractor unit is in bipolar in nature and it is to be converted in to unipolar video by using full wave rectifier before presenting on display unit.

By Practical MTI system is capable of cancelling echoes from fixed targets having amplitude of the order of 40 db or greater above visibility. At the same time it is possible to observe moving targets superimposed over the fixed targets and which are weaker than the fixed targets by as much as 25 db.

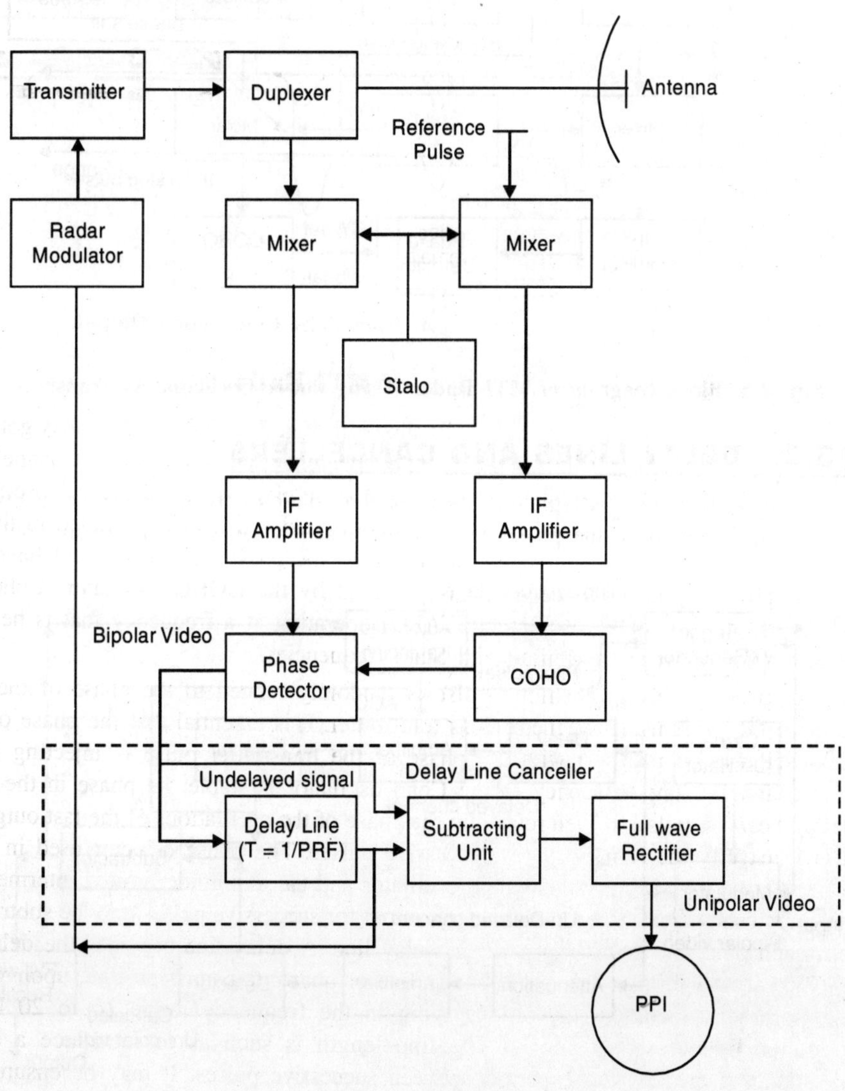

Fig. 3.2. Block Diagram of MTI Radar with the Delay Line Canceller.

3.2.2. MTI Radar using Power Oscillator as Transmitter

A block diagram of MTI radar using a power oscillator is shown in Fig. 3.3. A portion of the transmitted signal is mixed with the STALO output to produce an IF beat signal whose phase is directly related to the phase of the transmitter. This IF pulse is applied to the coherent oscillator (COHO) and causes the phase of the COHO CW oscillation to "lock" in step with the phase of the IF reference pulse. The phase of the COHO is then related to the phase of the transmitted pulse and may be used as the reference signal for echoes received from the particular transmitted pulse. Upon the next transmission another IF locking pulse is generated relock the phase of CW COHO until the next locking pulse comes along. This type of MTI radar has wide applications.

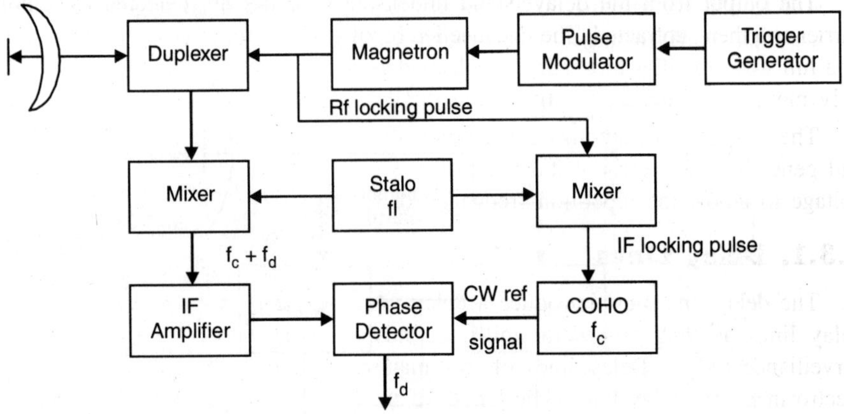

Fig. 3.3. Block Diagram of MTI Radar (Using Power Oscillator As Transmitter).

3.3. DELAY LINES AND CANCELLERS

A block diagram of a typical delay lines cancellation network is shown in Fig. 3.4. The bipolar video from the phase detector modulates a carrier before being applied to

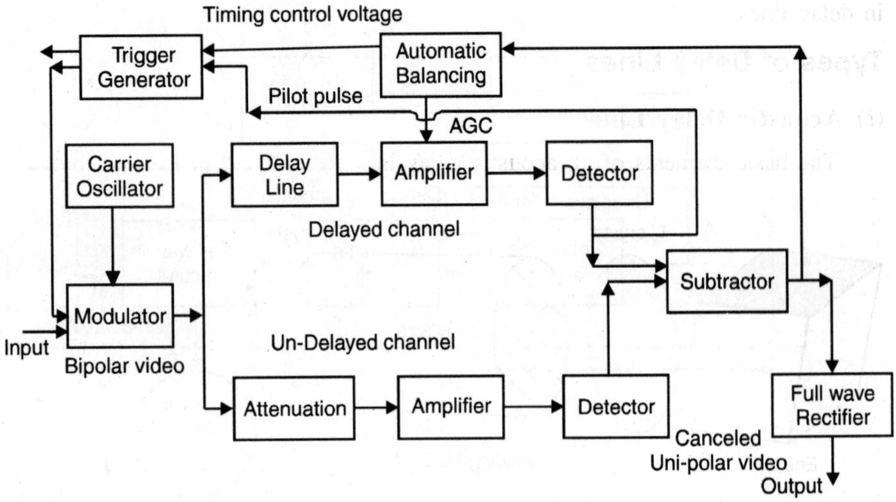

Fig. 3.4. A Block Diagram of Single Delay Line Canceller.

the delay lines. The carrier frequency is normally 15 or 30 Mhz. The radar output is not directly applied to the delay lines as a video signal since it would be differentiated by the crystal transducer that convert the electromagnetic energy into acoustic energy, and vice versa. The modulated bipolar video is divided between two-channel. In one channel the signal is delayed by a PRP, while in the other channel it reaches directly *i.e.*, undelayed. There is considerable attenuation in the signal introduced by the delay line and must be amplified in order to bring it back to its original level. Since the introduction of an amplifier into the delay channel can alter the phase of the delayed waveforms and introduce a line delay, an amplifier with the same delay characteristic is also used in the direct (undelayed) channel. An attenuator might also be inserted in the direct channel to make equalizing the gain. Good cancellation in a typical application might result in an uncancelled voltage residue of the order of 1% or 40 dB.

The output from the delayed and undelayed channels are detected to remove the carrier and then subtracted. The uncancelled bipolar video from the sub tractor is rectified in a full wave rectifier to obtain unipolar video signal for displaying on the PPI. Ideally, only moving produced an output from the subtractor.

The purpose of automatic balancing to detect any amplitude timing differences and generates AGC error voltage to adjust the amplifier gain and timing control error voltage to adjust the repetition frequency of the trigger generator.

3.3.1. Delay Lines

The delay line must introduce a delay equal to the pulse–repetition interval. Thus delay lines as long as several milli seconds are required for typical ground based surveillance radars. Delay lines of this magnitude cannot be achieved with practical electro magnetic delay lines. The length of the EM delay path would have to be equal to twice the unambiguous range of the radar. Converting the EM wave in to acoustic waves and accomplishing the delay in an acoustical delay line overcome this difficulty. After the delay the acoustic wave are converted back to EM waves. The velocity of acoustic wave depends on the delay medium, but it is of the order of magnitude of 10^{-5} that of EM waves; both liquids and solids have been used as the acoustic delay media. Mercury and fused quartz are the two media most widely employed for MTI radar application, but water, water–glycol mixture, aluminium, magnesium and glass have also been used in delay lines.

Types of Delay Lines

(*i*) Acoustic Delay Line

The basic elements of an acoustic delay line are outlined in Fig. 3.5 below.

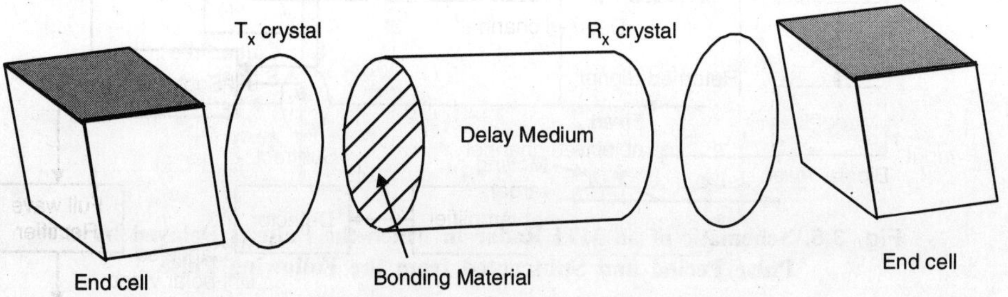

Fig. 3.5. Elements of an Acoustic Delay Line.

The EM energy is converted into acoustic energy by piezoelectric transmitting crystal. A similar transducer (the receiving crystal) at the o/p of the line converts the acoustic energy back to EM energy.

(*ii*) Quartz Crystal

The **Quartz crystal** transducers are normally a high Q device with an inherently small bandwidth. However, when transducer is coupled to a delay medium, the medium has a dumping effect, which broadens the bandwidth. Consequently, acoustic delay lines are relatively broadband devices.

(*iii*) Liquid Mercury

One of the simplest acoustic delay lines consists of a straight cylindrical tube filled with **liquid mercury**. The transit time of acoustic waves in mercury at room temperature is approximately 17.5 μs/inch. To produce a delay of 1000 μs the line must be 57 inch in length exclusive of end cells. This is manageable size in ground-based radar.

A more compact configuration may be had by **folding the line** back on it itself one or more times. Another method of obtaining a more compact delay line is to make use of **multiple reflection** in a tank filled with liquid. The alignment of the reflecting surface is a problem, and it has been difficult to obtain a leak proof construction.

One of the disadvantages of either solid or liquid delay is the large insertion loss. The insertion loss of a typical folded mercury delay line with 1000 μ sec is as follows:

Impedance mismatch loss of the crystal	36 dB
Free space attenuation in mercury	12.8 dB
Tubular attenuation	2.7 dB

The total insertion loss is 52 dB.

3.3.1.1. Modern Trends in Delay Line using Fiber Optics in MTI Radar

In the advanced version of the radar, nowadays we are using Fiber Optics as a delay-line media, which is an excellent propagating medium for time delays. It has low loss, low dispersion, lightweight, and is inexpensive. (Details of this is beyond the scope of this book). (Reference : IEEE Transaction on Microwave theory and techniques, Vol. 45, No. 8 August 1997-1473).

In MTI radar, the small echo from the target has to be distinguished from the large echo from fixed objects (clutter). The filtering of the clutter in a typical MTI radar system is performed by delaying a returned pulse by the pulse period, and subtracting it from the following pulse, as shown in Fig. 3.6.

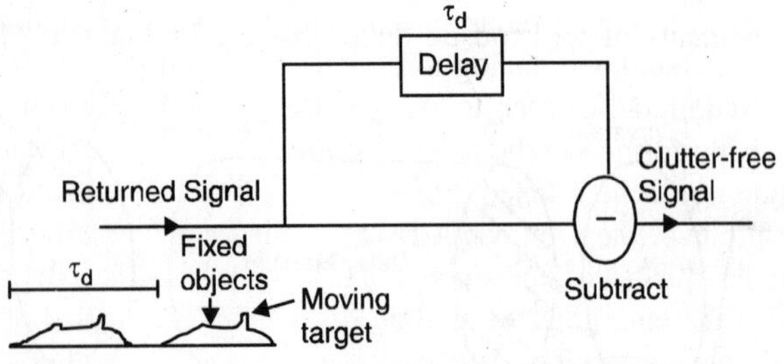

Fig. 3.6. Schematic of an MTI Radar in which the Pulse is Delayed by the Pulse Period and Subtracted from the Following Pulse.

Signal processing using optic-fiber delay lines is a very powerful technique for processing high bandwidth signals. Moreover, the signal carried by the intensity modulated optical beam is insensitive to electromagnetic radiation. Various optical delay-line configurations have been proposed to achieve different radio-frequency (RF) filter functions. These configurations can be either optically coherent or incoherent. Coherent optical processing allows more flexibility in shaping of the filter-transfer function. Nevertheless, the phase of the optical beam has to be precisely controlled since the filter is highly sensitive to any phase variation. Moreover, the longest time delay in the filter has to be shorter than the coherence time of the optical source. Incoherent processing does not require any control of the optical phase. The filter is, therefore, less sensitive to external perturbation (*e.g.*, variation of temperature). Long time delays are made possible, but the shortest delay line has to be longer than the coherence length of the optical source. In this approach, optical signals are detected using a pair of matched photo detectors and combined electrically. Another unique advantage of the filter is that the two signals, which recombine inside the photo detector, are at two different wavelengths avoiding problems of optical coherence, which appear in traditional delay-line filters.

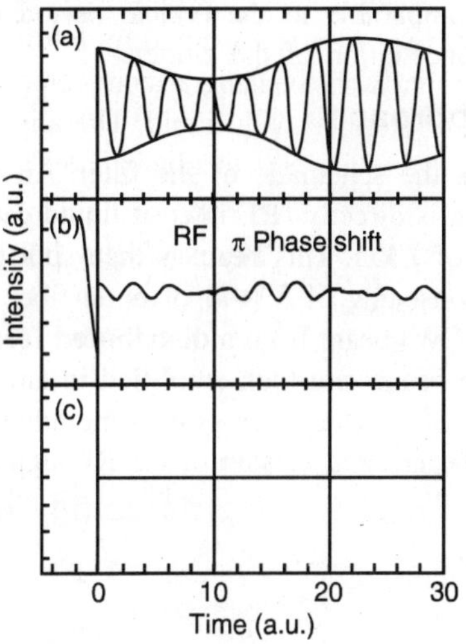

Fig. 3.7.

(*a*) Time-domain representation of a radar pulse. (*b*) Resulting filtered signal after filtering using π phase shift of an RF carrier—due to the π phase shift the cancellation is incomplete. (*c*) Resulting filtered signal after filtering using inversion inside an semiconductor operational amplifier (SOA)—the cancellation is theoretically complete.

A major application of the feed forward filter is in radar technology. In MTI radar, the small echo from the target has to be distinguished from the large echo from fixed objects (clutter). In a typical radar system, this filtering is digitally realized after down conversion to base band and analog-to-digital (A/D) conversion. Typically, 14–18 dB of resolution are required in order to resolve the faint target from the strong clutter. If some filtering is performed in the analog domain before A/D conversion, the requirement on A/D resolution will be relaxed. For example, with a 30-dB attenuation of the clutter in the analog domain, the required A/D resolution is reduced by 5 b. Optical filters are ideal to perform this filtering function at the front end of the antenna as they are compatible with the emerging technology of optically controlled phased-array antennas, and long low cost time delays are easily achieved. The clutter, which has the same phase from pulse-to-pulse, will be removed, while the doppler shifted-target echo will be preserved. All-optical subtraction is, therefore, a key operation of the filtering function. In principle, subtraction in the analog domain can also be made by shifting one of the signals by half the RF period out of phase) and then combining them. However, for

MTI radar this technique does not lead to a perfect cancellation of the clutter as analytically shown in Fig. 3.7(*a*) represents the simulated pulse of a radar, which consists of an RF signal amplitude modulated. Fig. 3.7(*c*) shows the filtered signal after recombination of the signal represented in Fig. 3.7(*a*) with a RF phase shifted version of itself. Although strong attenuations achieved, part of the signal still remains. The no perfect cancellation stems from the fact that a phase shift is not necessarily equivalent to an inversion. In contrast, Fig. 3.7(*c*) shows the cancellation when the signal represented in Fig. 3.7(*a*) is recombined with an inverted version of itself. A theoretically perfect cancellation is obtained. The difference between these two techniques will depend on the relative frequency of the RF signal compared to the slow modulation on top of the return pulse. At high RF frequencies, the small extra time delay introduced to perform the phase shift will become negligible compared to the typical period of the intensity modulation. Therefore, the difference between inversion and phase shift will not be significant. However, at lower frequencies, the extra time delay introduced to perform the shift may become comparable to the typical period of the intensity modulation, leading to a non-ideal cancellation of the clutter.

All-Optical Filter Approach

In Fig. 3.8 describes the schematic of the filter. An external cavity tunable laser operating at λ_1 =1525 nm is directly RF intensity modulated. The coherence length of the laser is on the order of 1 km. This beam is then split into two parts; one part goes to the long time delay consisting of 500 m of optic fiber. The other part is combined with a continuous wave (CW) beam from a **distributed feedback (DFB)** laser operating at λ_2 =1526.6 nm. These beams are then amplified in an erbium doped fiber amplifier (EDFA) and sent into the SOA. Due to cross-gain modulation in the SOA, the emerging signal λ_2 at represents the inverted version of the RF signal.

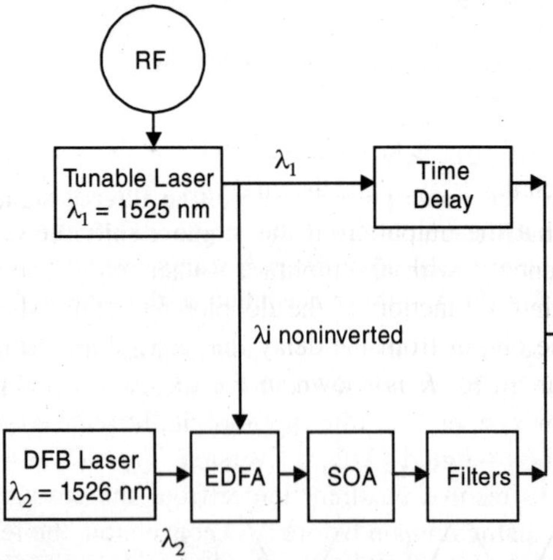

Fig. 3.8. Schematic representation of the filter—a directly modulated laser beam is split into two parts. One part is combined with a CW beam at λ_2 both beams are then amplified in an EDFA and sent to an SOA. The negative tap at λ_2 is then combined with the delayed positive tap at λ_1 and detected inside the same photo detector.

Using a differential configuration of the filter as shown in Fig. 3.8. This all-optical notch filter can be used in MTI technology to filter out the clutter signal. It can be used to filter the clutter at Intermediate frequency (IF) in the analog domain prior to A/D conversion, thus alleviating the requirement for high-resolution A/D converters.

Moreover, as the signal stays in the optical domain, cascading of delay-line processors is made possible.

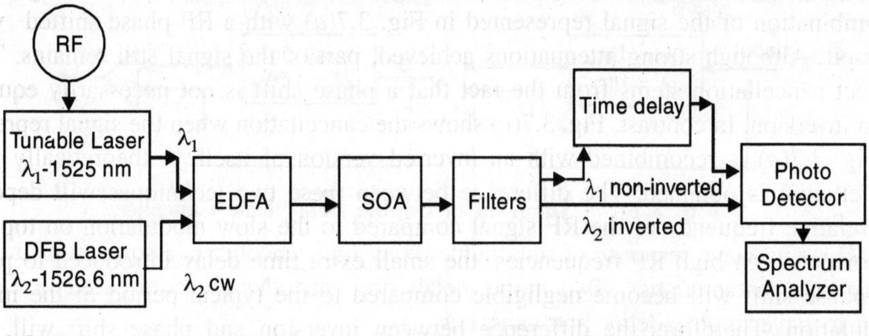

Fig. 3.9. Schematic representation of the differential configuration of the filter, showing improved performance. Beams at λ_1 and λ_2 go through the SOA. The positive tap at λ_1 is delayed and is recombined with the negative tap at λ_2. Both beams are then detected in the same photo detector.

3.3.1.2. Response of The Delay Lines Cancellers

The video signal received from a particular target at a range *R* can be expressed as,

$$V_1 = K \sin (2\pi f_d t - \phi_0)$$

Where *K* is the amplitude of video signal and ϕ_0 =Phase shift ($4\pi ft\ R/c$)

At a time *t-T*, where *T* is the Pulse repetition time, the voltage from the same target is

$$V_2 = K \sin (2\pi f_d(t - T) - \phi_0$$

The output from the subtractor is given below, provided rest every thing remain constant over the time interval *T*.

$$V = V_1/2 - V_2/2 = K \sin \pi f_d T \cos \{2\pi f_d(t - T/2) - \phi_0\} \quad ...(3.1)$$

The each video voltage is divided by the factor ½ because it is assumed that voltage of video, are equally divided between the delayed and the un delayed channel of the receiver.

Above Eqn. (3.1) indicates that the output from the single–delay line canceller is a cosine wave at a doppler frequency f_d with an amplitude $2K \sin \pi f_d T$. The amplitude of the canceller video output is thus a function of the doppler frequency shift and the PRT. The ratio of amplitude of the output from the delay line canceller, $K \sin (\pi f_d T)$ to the amplitude of the normal radar video, *K* is shown in Fig. 3.10.

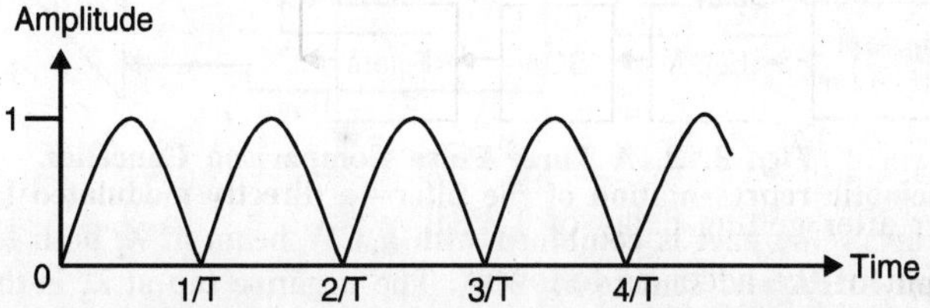

Fig. 3.10. Frequency Response of the Single Delay Canceller.

3.3.1.3. Double Cancellation

A single delay line canceller is not sufficient to reject all the clutter as desired in the vicinity of DC components or at doppler frequencies corresponding to the different

PRF and its harmonics. The frequency response function of double delay line canceller is much wider than the single delay line canceller. To have the widened clutter rejection, the output must be pass through the second delay line canceller.

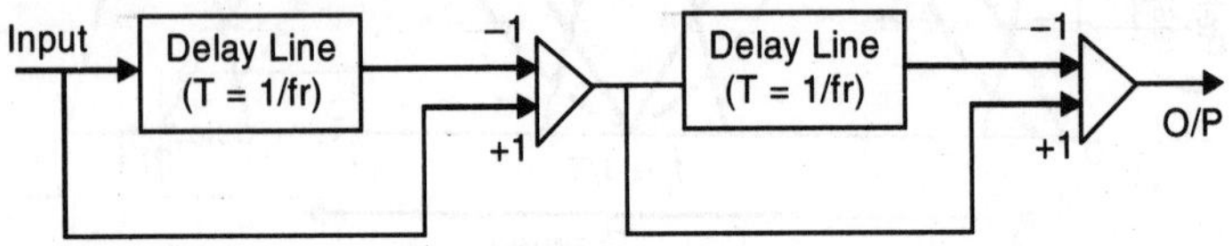

Fig. 3.11. A Diagram of Double Delay Line Canceller.

The sine wave response of the double delay line canceller is a squared of the single delay line characteristics. Two single delay line canceller are configured in cascaded form to have double delay line canceller.

A double delay line canceller shown in the Fig. 3.11. Two delay lines are being arranged in the cascaded form alongwith the two adders. A input signal $f(t)$ is inserted in the Ist delay line canceller and to Ist adder simultaneously. The output from the Ist delay line canceller after delaying of one PRF ($T = 1/f_r$) feed to the Ist adder. In the next stage, the output of adder is bifurcated and one is given to the IInd adder directly and the second input is feed through the IInd delay line canceller. The output of double delay canceller is as follows :

$$f(t) - f(t + T) - f(t + T) + f(t + 2T) \qquad ...(3.2)$$

There is one more configuration commonly known as the three pulse comparison as shown in Fig. 3.12. A three-pulse comparison canceller configuration has the same frequency response characteristics as the double delay canceller. The operation of this configuration is explained here. In this configuration, we use two delay line canceller and a adder. A input signal $f(t)$ is bifurcated and one is fed to the Ist delay line canceller and second is given to the adder, which is connected at the output end of IInd delay line canceller. The output from the Ist delay line canceller (after introducing delay of 1 PRP) is again divided in two parts and one is fed to the IInd delay line canceller and other part is given to the adder directly. The output of IInd delay line canceller is finally fed to the adder. The adder get three signal, one is directly from the input signal, second from the IInd delay line canceller after getting delay of 2 PRP and the third input is from the output end of 1st delay line.

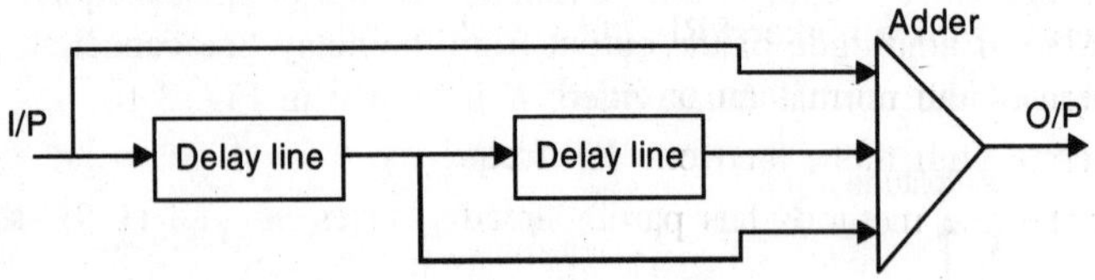

Fig. 3.12. A Three Pulse Comparison Canceller.

Canceller after getting delay of 1 PRP.

The output of the adder as follows :

$$f(t) - 2f(t + T) + f(t + 2T) \qquad ...(3.3)$$

The three-pulse comparison canceller is same as the double delay canceller until in pefect adjustment. The frequency response characteristics curve is shown in the Fig. 3.13.

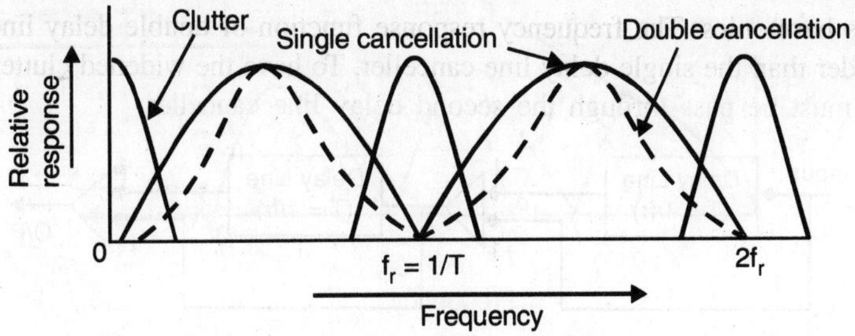

Fig. 3.13. Frequency Response of Single Delay Line Canceller and Double Delay Line Canceller.

3.3.1.4. Blind Speed

Blind speed is a very serious problem in MTI radar. When the doppler shift equals an integer multiple of PRF (Pulse Repetition Frequency), the delay line canceller not only eliminate the DC components caused by clutter but it also rejects moving targets. In other words, blind speed can be explained as the relative velocities at which the MTI response is zero called **blind speeds.**

The response from the single delay line canceller will be zero, When

$$f_d = n/T = nf_r$$

Where $n = 0, 1, 2, 3, ...; T = PRT$ and f_r is the PRF.

The nth blind speed V_n can be expressed as

$$V_n = n\lambda/2\ T = n\lambda f_r/2$$

Where, $n = 1, 2, 3, ...$

If λ is in meters, f_r in hz and the relative velocity in knots.

The blind speeds are written as

$$V_n = n\lambda f_r / 1.02 = n\lambda f_r \qquad ...(3.4)$$

This limitation of blind speed can be reduced up to some extent by employing following steps :

1. Operate the radar at long wavelength (means low frequency).
2. Operate with a higher PRF.
3. Operate with more than one PRF.
4. Operate with more then one RF frequency.

Choice of these methods has particular advantages as well as limitations.

3.4. STAGGERED PULSE REPETITION FREQUENCIES

If a radar is operating at multiples PRFs or its PRF is changed either pulse-to-pulse or scan-to-scan, than the affect of blind speed can be eliminated from the radar. If two radar operating at same frequencies but having its different PRF then if one radar is blind to moving target. So, if we use single radar but having different PRF than the same affect can be achieved. When the PRF is changing pulse to pulse than it may be called as staggered PRF. Staggering of PRF is generally employed in Air Traffic Control Radar such as Surveillance Radar Element (SRE).

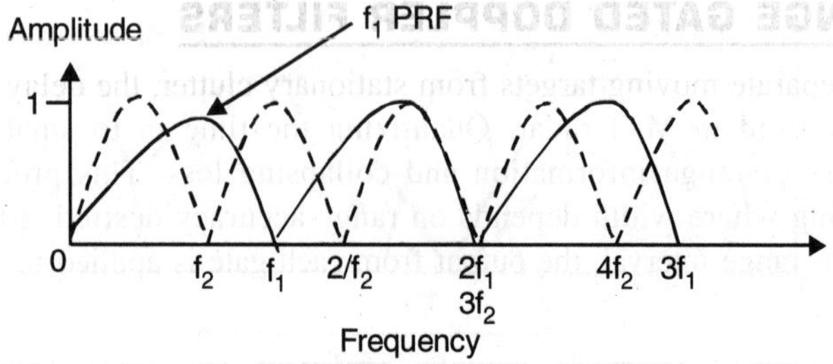

Fig. 3.14. Frequency Response of Two PRFs F_1 and F_2, where $F_1 = 3F_2/2$.

In the Fig. 3.14, frequency response of two PRF is shown. Suppose the first PRF is F_1 shown in bold line and the blind speed of second PRF is F_2 shown in dotted lines. If we observe the Fig. 3.14, it is clear that at particular position when $2f_1 = 3f_2$, both the PRFs have the same blind speed.

The multiples PRFs can be obtained by using several methods. Using following techniques may vary the PRFs :

1. Pulse-to-pulse (known as staggered PRF)
2. Scan-to-scan
3. Dwell-to-dwell.

The problems occurs in using staggered PRF is that residual of uncancelled echoes of clutters, which are due to second time around echoes. So to minimize the second time around echoes affect, if we use unstaggered PRF in the sector where second time around are expected more and rest of the sector used staggered PRFs.

3.4.1. Sub Clutter Visibility

Sub clutter visibility may be defined as the ability of MTI radar to detect the moving target, if the target is superimposed over the clutters. SCV defined the performance of an MTI radar and it is measured in dBs. It may be defined as

$$\text{SCV} = \text{The gain in signal-to-clutter power ratio.}$$

Suppose an MTI radar has SCV 20 db than it means that a moving target can be detected in the presence of clutter even though the clutter echo power is 100 times the target echo power.

3.4.2. The Cancellation Ratio

The cancellation ratio may be defined as the ratio of a fixed target signal voltage using MTI cancellation to the signal voltage without MTI cancellation. We may write

$$\text{Cancellation ratio (CR)} = \frac{\text{Fixed targets signal voltage after MTI cancellation}}{\text{Signal voltage without MTI cancellation}}$$

The CR is a number that is always less than One (1) and may be expressed in db.

3.4.3. The Target Visibility Factor

The target visibility factor is the ratio of the signal strength from a target, which is moving at a specified radial velocity to the signal strength from the same target when it is moving at an optimum radial velocity.

3.5. RANGE GATED DOPPLER FILTERS

In order to separate moving targets from stationary clutter, the delay line canceller has been widely used in MTI radar. Quantizing the time in to small interval can eliminate the loss of range information and collapsing loss. This process is known as the range gating where width depends on range accuracy desired. After quantizing the radar return to range interval, the output from each gate is applied to a narrow band filter.

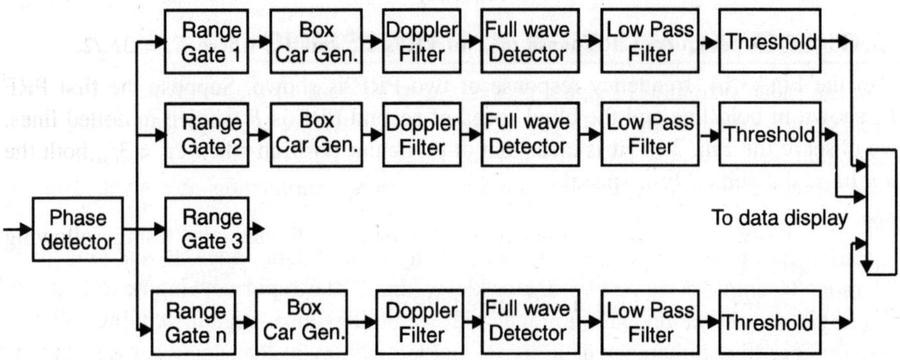

Fig. 3.15. Block Diagram of MTI Radar Using Range Gates and Doppler Filters.

A block diagram of the video of an MTI radar using multiple range gates followed by clutter rejection filter is shown in Fig. 3.15. Here the range gates sample the output of the phase detector sequentially range interval. Each range open in sequence just long enough to sample the voltage of the video waveform corresponding to a different range interval in space or it acts as a switch/gate which open and close at the proper time. The output of the range gate is given to a circuit known as box car generator. Its function is to aid in the filtering and detection process enhancing the fundamental of the modulation frequency and eliminating harmonics of the PRF. The clutter rejection filter is nothing but a band pass filter whose bandwidth depends on the extent of the expected clutter spectrum. The filtered output from the Doppler filter is further feed to a full wave linear detector which convert the bipolar video to unipolar video. A low pass filter or integrator passes these unipolar video to the threshold detection circuit. Any signal crosses the threshold level is treated as a target. The outputs from each range channels are combined for display on the PPI or any other display unit.

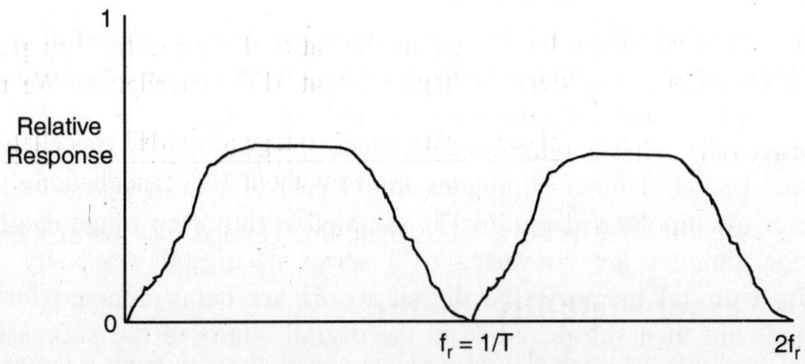

Fig. 3.16. Frequency Response Characteristics of an MTI using Range Gates and Dopplers Filters.

The presentation of this type of MTI radar is far better than the display from normal MTI radar.

The frequency response characteristics of an MTI radar using range gates and filter is shown in Fig. 3.16, the shape of the rejection band is mainly determined by the shape of the band pass filter.

It must be pointed out that the MTI radar using range gates and filters is more complex then an MTI with single delay line canceller a better MTI performance is achieved from a better match between clutter filter characteristics and clutter spectrum.

3.6. DIGITAL MTI SIGNAL PROCESSING

In the radar system MTI signal processing to be done in the receiver. The received echoes signals either from the target or from the clutter are to be processed in such a manner that at the display unit only moving targets to be available. The signal strength should be good enough to display the target without introducing any noise over the scope.

Earlier, in an MTI processing we were using analog delay lines, it was rare for an MTI radar to employs more two analog delay lines. The rapid development of digital technology brought the revolution in the radar technology also. Digital technique allowed radar designer to use memory as a digital storage device, in that delay can be obtained for whatever length of time was required.

The advantages offered by digital MTI process include removing of blind phase in MTI radar by using *I* and *Q* channel. When the phase between the Doppler signal and the sampling at the PRF results in a loss, it is called a blind phase. It is different from the blind speed.

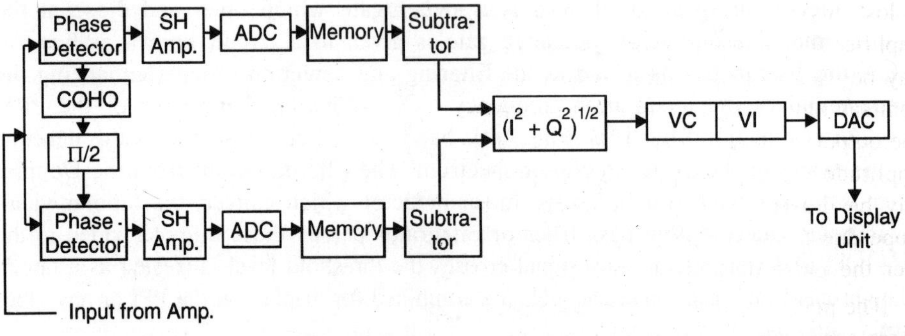

Fig. 3.17. A Block Diagram of a Digital MTI Signal Processor.

A block diagram of digital MTI Doppler signal processor is shown in the Fig. 3.17. The output from the IF amplifier is split in to two channels, one for in–phase channel denoted "*I*" and the other for quadrature channel denoted by "*Q*" and given to the respective phase detector. The outputs of the phase detectors are 90 degree out of phase to each other. The *Q* channel eliminates the effects of blind phases. Following the phase detector the bipolar video signal is sampled within each range resolution cell. These voltage samples are converted to a series of digital words by using A/D converter. In a digital memory, the digital words are being delayed for one PRP. Further signals are then subtracted from the digital words of the successive sweep. By taking the square root of $(I^2 + Q^2)$, The digital outputs of the *I* and *Q* channels are combined. This also can be done alternatively, which is considered adequate by taking $|I| + |Q|$. The combined unipolar output may be further processed for optimization of the signal. Further processing of these signals may be such as video correlator, video integrator.

3.7. NON-COHERENT MTI

The echo signal received from a moving target or from clutter fluctuates both in amplitude and phase. Where the MTI makes use of the phase fluctuation than it is called coherent MTI and where the amplitude fluctuation is being considered than it is called non-coherent MTI. In non-coherent MTI, the amplitude fluctuation is used to recognize the doppler components produced by a moving target. It is also called externally coherent.

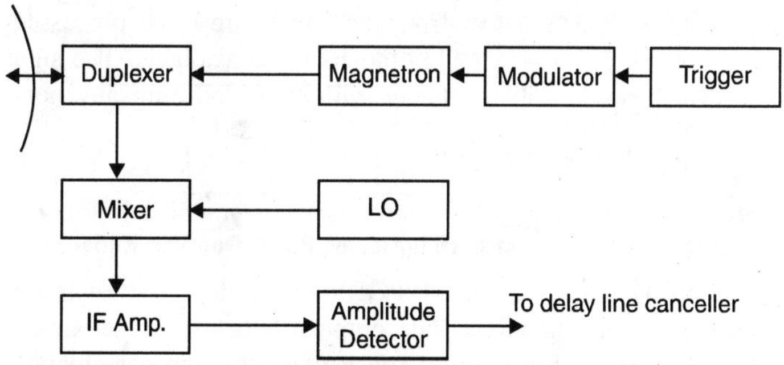

Fig. 3.18. Block Diagram of Non-Coherent MTI Radar.

The block diagram of non-coherent MTI is shown as Fig. 3.18. In non-coherent MTI amplitude limiter can not be used otherwise desired amplitude fluctuation would be lost. Instead of using phase detector we are using amplitude detector. Therefore, IF amplifier should be linear and should have large dynamic range. A logarithm amplifier may be used as IF amplifier to have logarithm gain characteristics, such as protection from saturation and have uniform output with variations in the clutter input amplitude. The output of IF amplifier to be detected through the amplitude detector. The output of amplitude detector may be further processed either using delay line canceller or it may be detected over the A-scope. A butterfly effect can be observed on the A-scope due to the doppler in amplitude fluctuation. The Transmitter should be stable over the pulse duration to prevent beat from the overlapping ground clutter.

The main advantage of non-coherent MTI is that it is very simple and may be used where space and weight are limited.

3.8. PULSE DOPPLER RADAR

Pulse radar is a combination of pulse radar and CW radar. It works on the principal of Doppler shift as MTI radar follows. As per the Nyquist Criterion the sampling rate (*i.e.* PRF) should be greater and equal to the twice of the doppler shift frequency but in MTI due to use of low frequency its became under sampled. It will leads to ambiguous estimation of target speed and occurrence of blind speed, where target appears stationary and unresolvable against back ground clutter. Pulse Doppler radar being high PRF radar, it can remove the doppler ambiguities. To extract the doppler shift information of the carrier the pulse radar should be modified by introducing a coherent oscillator (COHO) for frequency stability in he transmitter and receiver chain. It employs the coherent radar system.

Pulse Doppler radar is classified as high PRF and as medium PRF. In high PRF pulse radar there is ambiguity in the range but unambiguities in the velocity. In the medium PRF pulse radar there is ambiguities in range and velocity both.

The block diagram of Pulse Doppler Radar is shown in Fig. 3.19.

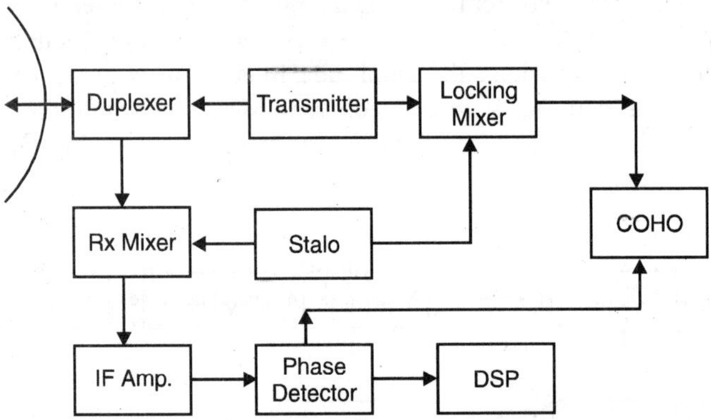

Fig. 3.19. The Block Diagram of Pulse Doppler Radar.

A STALO (stable local oscillator) is used to allow the phase of transmitter signal to be maintained by a locking mixer. The output of locking mixer given to lock the COHO phase and in urn it serves as reference phase for the detector at intermediate frequency. Now the, phase detector measures the difference in phase between two RF signal. Due to the target motion the phase path of the echo changes pulse to pulse and by the same amount phase difference will vary.

Applications. Pulse radar has many applications; some of those are listed below:

1. It is being used as weather warning radar at the airbases to detect and measure thunderstorm, turbulence in the air.
2. It is very useful in detecting and estimating the target motion, locking of particular target out of a group.
3. To observe thunderstorm, rain and hail, a double polarization doppler radar is being used.

Advantages :

1. A pulse doppler radar has got the ability to reject the unwanted echoes by using doppler filters or by a range gating.
2. It can measure the range and velocity over predetermined limits, even in presence of multiples target.
3. Signal to noise ratio can be increased by using coherent integration.

3.9. MTI FROM MOVING PLATFORM (AMTI)

When the radar is mounted on a ship or an aircraft and it is in motion, the detection of a moving target in presence of clutter becomes more difficult than when it is stationary. The doppler frequency shift of the clutter varies with the direction of antenna in azimuth and the elevation angle to the clutter, and clutter appears to be in motion. The design of an MTI is more critical with airborne radar in comparison to a ship borne radar due to the higher speed and the greater range of elevation angle.

An MTI radar when used on a moving platform is called AMTI radar where the letter "A" stood for Airborne. In airborne surveillance radar, the relative clutter velocity depends on aircraft velocity and the direction of the clutter relative to the aircraft velocity vector. In addition to the shift of the center frequency of the clutter, its spectrum is also widened. The spectral width can be obtained approximately by considering the differential of the Doppler frequency.

$$f_d = 2(v/\lambda)\cos\theta \quad ...(3.5)$$

or

$$\Delta f_d = 2v/\lambda \sin\theta\, \Delta\theta \quad ...(3.6)$$

where v = platforms speed
λ = wavelength
θ = Azimuth angle

If the beam width is taken as $\Delta\theta$, then Δf_d represents a measure of the width of doppler frequency spectrum. The effect of platform velocity may be considered as having two components, one is in the direction of antenna pointing and the other is normal to the direction of antenna pointing.

3.9.1. Coherent AMTI

However, in general all MTI techniques can be applied to the AMTI but it is usually more difficult to achieve good MTI radar.

The coherent MTI radar may be used as AMTI radar. The frequency of the Coho is shifted to compensate for the relative velocity of the radar platform with respect to the clutter. The block diagram of coherent AMTI is shown as Fig. 3.20. In this, a doppler-frequency oscillator is being used, which is a tuned oscillator. The output of this oscillator is made to be proportional to the relative velocity between radar and clutter and may be controlled according to the position of the antenna with respect to clutter. The output of the Coho is mixed with a signal from the doppler frequency oscillator. Instead of using Coho as the references to the phase detector, one of the side bands of the heterodyned, signal is selected by a narrow band filter and is used for references.

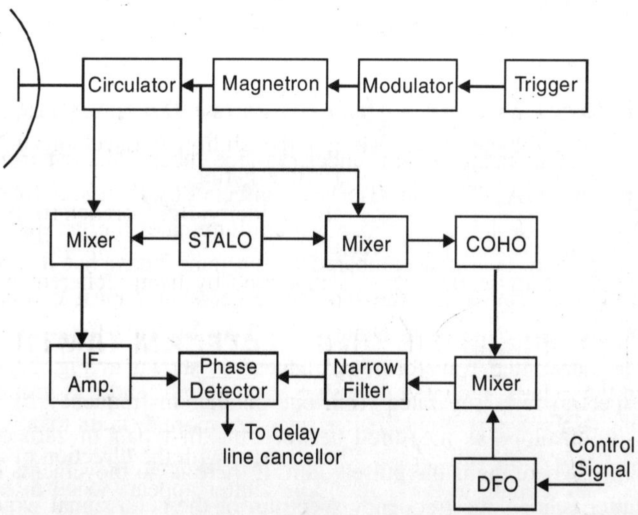

Fig. 3.20. Block Diagram of Coherent AMTI Radar.

The signal is coherent with the transmitted signal but it is shifted in frequency velocity of clutter.

The frequency of the Doppler compensation oscillator must be correspondingly changed since the relative clutter velocity changes with the direction of the antenna beam. Therefore the effective doppler frequency of the ground clutter may vary appreciable over the range interval of interest and make the doppler compensation of the Coho signal extremely difficult.

3.9.2. Pulse Doppler AMTI

A pulse Doppler MTI radar can be a better form of AMTI radar. In this using a rejecter filter can eliminate the ground clutter signal, which are being shifted in frequency by the doppler effect. If the rejection filter cannot continuously track the changing doppler frequency caused by a relative velocity, a narrow pencil beam may be used in which changes in Doppler occur as antenna is scanned in angle.

3.9.3. Non-Coherent AMTI

Due to less weight and space occupied by a non-coherent MTI, it is being preferred in aircrafts. The non-coherent AMTI is limited, as its ground based counter part, by the need for sufficient clutter signal to provide the references upon which the doppler fluctuations may be detected.

3.9.4. Fluctuation Caused by Platform Motion

In addition of the limitations of the MTI radar, AMTI radar has another serious problem of fluctuation caused by motion of radar platform. This effect is quite same as antenna fluctuation. The clutter that the radar illuminates consists of number of independent scatters randomly and located within the resolution cell of the radar. The echo signal from each of these scatter add vectorically at the receiving antenna. As the radar beam moves between pulses, the distance to each of the scatters changes. A change in distance result in a change in phase, and the vector addition of the echo signal from all the scatters may not be the same from pulse-to-pulse. Thus the clutter return will fluctuate from pulse-to-pulse and an uncancelled residue will result at the output at the delay line canceller.

3.9.5. Effects of Side Lobes on Pulse Doppler AMTI Radar

As in the other radar system, undesired side lobes radiation from an antenna is a serious problem, in AMTI radar also being affected by their side lobes. In AMTI radar the trouble some side lobes are those, which illuminates the ground. Although the side lobes radiators may be small compared to the main beam but it will contribute large clutters from the ground. So the moving target signal must compete not only with clutter illuminated by the main beam as well by the side lobes too. The spectrum of pulse radar transmitted waveform will appear as shown in Fig. 3.21. It consists of a series of spectral lines separated from one another in frequency by the PRF (f_r). The width of the envelope as measured between the first pair of zero closing about f_0 is equal to $2/pw$, where pw is the pulse width. If there is no movements in target and radar and no clutter echoes, the frequency spectrum of the echo signal would be the same as that of the transmitted signal. However, the relative motion between radar and target as well as between radar and clutter and the additional clutter signal received from the antenna side lobes will modify this signal spectrum.

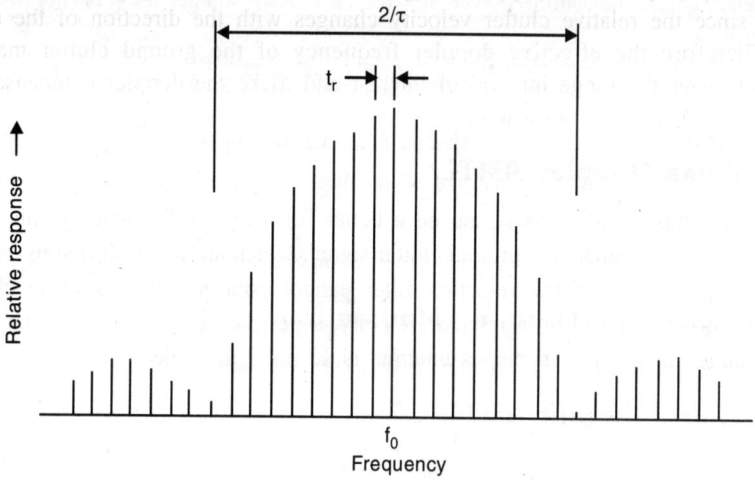

Fig. 3.21. **Spectrum of Pulse-Radar Transmitted Waveform.**

3.10. LIMITATIONS TO MTI PERFORMANCE

An improvement in the signal/clutter ratio of an MTI is affected by several factors other than the design of the doppler signal processor, equipment instabilities, internal fluctuation of clutter, scanning modulation and limiting in the receiver can all detract from the performance of an MTI radar.

3.10.1. Equipment Instabilities

Pulse-to-pulse changes in the amplitude, frequency or phase of the transmitter signal lower the improvement factor of an MTI radar. If the echo from stationary clutter on the first pulse is $A \cos \omega t$ and from the second pulse is $A \cos (\omega t + \Delta Ø)$, then the difference between the two

$$= A \cos \omega t - A \cos (\omega t + \Delta Ø)$$

where $\Delta Ø$ = change in oscillator phase (Ø).

$$= 2A \sin(\Delta Ø/2) \sin(\omega t + \Delta Ø/2)$$

for small phase errors, the amplitude of the resultant difference

$$2A \sin (\Delta Ø/2) = A \Delta Ø$$

So the limitation on the improvement factor due to oscillator instability is

$$I = 1/(\Delta Ø)^2 \quad ...(3.7)$$

This would apply to the coho locking or to the phase change, which is introduced by a power amplifier.

3.10.2. Internal Fluctuation of Clutter

There are many types of clutter which are not absolutely stationary like that due to buildings, water towers, hills, mountains etc. Echoes from rain, sea, trees, vegetarian etc, fluctuate with time and there limit the performance of MTI radar. Most of the fluctuation clutter targets can be represented by a model consisting of many independent scatters situated within the resolution cell of the radar.

Experimentally measured power spectra of clutter signals may be approximately written as

$$W(f) = |G(f)|^2 = |g_0|^2 \exp[-a(f/f_0)^2] \quad ...(3.8)$$

where $W(f)$ = clutter power spectrum as a function of frequency.

$G(f)$ = Fourier transform of the input waveform.

f_0 = radar carrier frequency.

a = Parameter which depends on the clutter.

3.10.3. Antenna Scanning Modulation

As the antenna scans by a target, it observes the targets for a finite time 't_0'

Where $$t_0 = n_B / f_r = \theta_B/\theta_s$$

n_B = number of hits received

f_r = PRF

θ_B = antenna beam width

θ_s = antenna scanning rate.

The received pulse train of duration to have a frequency spectrum whose width is proportion to$1/t_0$. So even for a perfectly stationary clutter, there will be finite width of the clutter spectrum owing to the finite time on the target. When the clutter spectrum is too wide, it affects the improvement factor. This limitation is also called scanning modulation or scanning fluctuation.

The limitation to the improvement factor caused by antenna scanning are

$$I1s = n_B^2/1.380 \quad \text{(single canceller)}$$

$$I1s = n_B^4/3.852 \quad \text{(double canceller)}$$

3.10.4. Limiting in MTI Radar

Before the MTI processor, a limiter is generally employed in the IF amplifier for preventing the residue from large clutter echoes. An ideal MTI radar should reduce the clutter to a level comparable to receiver noise. If the limit level relative to noise is set higher than the improvement factor, clutter residue obscurer part of the display while if is set too low there may be a 'black hole' effect on the display. The limiter provides a CFAR and thus it serves a very essential part to obtain usable MTI performance. The use of the limiter eliminates the amplitude information of the IF output holding it constant to the limiting level and therefore, such an MTI radar may be called a phase processing MTI since only the phase information is retained after limiting.

3.10.5. Envelope Processing in MTI Radar

If the envelope of the IF output is detected by an envelope detector, the phase information of the output is of no use and such an MTI is, therefore also called an amplitude processing MTI, or in coherent MTI. An advantage of this system is that the local oscillator need not be as stable as in other system which employ synchronous or coherent detection. However, the disadvantage is that the signal to noise ratio with the envelope detector is less and, therefore, clutter must be present in relatively larger amount in the envelope processing MTI for the detection of a moving target, for, the clutter serves the same function as the reference signal in the coherent MTI, that employs a coherent oscillator (Coho).

3.11. MOVING TARGET DETECTOR (MTD)

The moving target detector (MTD) employs all the techniques to achieve a good overall moving target detection performance. The MTD is an enhanced configuration of moving target indicator (MTI) that combines a series of features to improve clutter rejection and target detection. The MTD term is applied to the Doppler processing system used in airport surveillance radars.

Utilization of primary radar data in the FAA automated air traffic control system has been impeded by the inability of existing signal processors to reliably detect aircraft in regions of strong ground and precipitation clutter, particularly when the aircraft are moving tangentially to the radar. During the past four years, under FAA sponsorship, M.I.T. Lincoln Laboratory has developed new techniques which significantly enhance automated aircraft detection in all forms of clutter. These techniques are embodied in a digital signal processor called the Moving Target Detector (MTD). This processor has been integrated into the ARTS-III system at the National Aviation Facilities Experimental Center, Atlantic City, New Jersey (NAFEC) and has undergone careful testing during the summer of 1975.

The main features in an MTD are :

1. The MTI pre-canceller.
2. The Doppler filter bank.
3. Use of burst-to-burst PRF diversity.
4. Adaptive thresholding.
5. The clutter map.

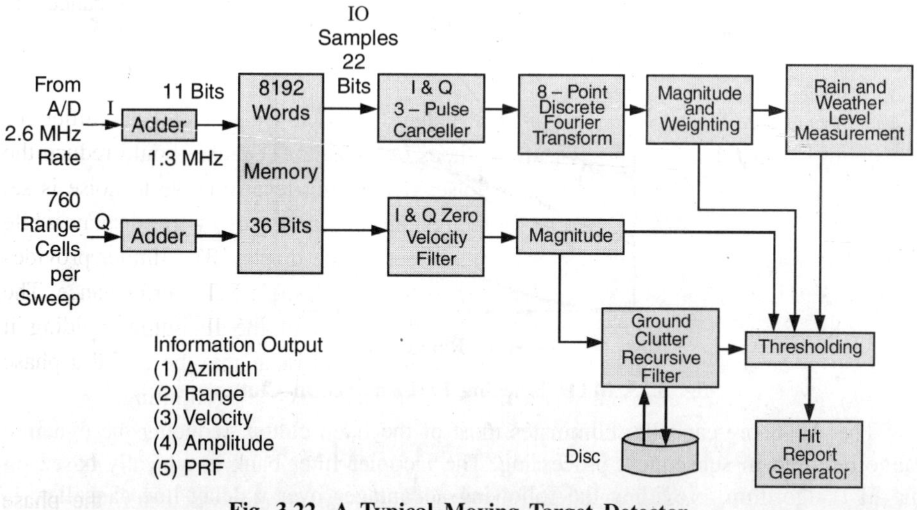

Fig. 3.22. A Typical Moving Target Detector.

A block diagram of the processor is presented in Fig. 3.22. The processor incorporates a digital ground clutter map which establishes the thresholds for the zero radial velocity cells. The clutter level in the ground clutter map continually adapts to a value based on the average level in the previous eight scans (32 seconds). Thresholding against ground clutter in this manner eliminates the usual MTI blind speed at zero radial velocity and permits the detection of tangential (zero radial velocity) aircraft over clutter if they have sufficiently large cross-section.

The non-zero radial velocity cells are threshold using the mean level of the signals in the same velocity filter averaged over one-half mile in range on either side of the cell of interest. The echoes from weather and/or receiver noise are averaged to establish a threshold. With this method of thresholding, the radar has subweather visibility; i.e., aircraft with radial velocity different than the rain can be seen even if their returns are substantially less than the total weather returns.

To eliminate blind speeds the pulse repetition rate is varied for each group of 10 pulses. This technique, when used with a coherent transmitter, also eliminates second-time-around ground clutter returns.

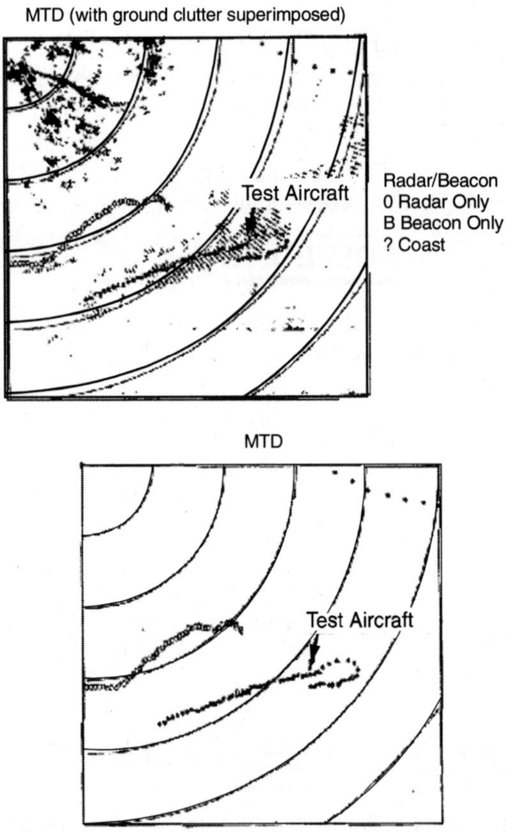

Two Miles Range Ring

Fig. 3.23. MTD Tracking Performance in Clutter.

The MTI pre-canceller eliminates most of the fixed clutter, reducing the dynamic range required in subsequent processing. The Doppler filter bank is typically based on the FFT algorithm, providing the following advantages over a delay line canceller:

1. The signal-to-noise ratio is improved by coherent integration within each of the n filters, whose bandwidth will be $1/n$ that of the canceller.
2. Doppler frequency measurement is available, based on the filter number in which detection occurs.

3. The filter bandwidth can be adjusted by amplitude or frequency weighting (windowing), giving better range side lobe reduction.
4. Adaptive thresholding can be applied to each filter, permitting rejection of moving clutter (weather clutter).

The use of burst-to-burst PRF diversity is necessary to fill blind speeds that fall within the target velocity region, and it provides a means of rejecting multiple time around (range ambiguous) target echoes. The adaptive thresholding applies separate thresholds to each filter, the nonzero velocity channels use a range cell averaging CFAR to adapt to moving clouds of precipitation, while the zero velocity channel threshold is generated by the clutter map, which applies a separate threshold for each range cell. In the zero velocity channel, which bypasses the MTI pre-canceller, targets at zero radial velocity and at the blind speeds can be detected if they exceed by a sufficient margin the clutter stored in the map.

The detection performance of the MTD was excellent in the clear, in rain and ground clutter, and false alarms were under complete control. The MTD processed range and azimuth data was very accurate, and the MTI did not suffer from track dropouts as did the conventional MTI when the aircraft track became tangential to the radar. Performance was excellent on magnetron as well as klystron-type radars with the exceptio-of second-time-around clutter cancellation.

SOLVED EXAMPLES

Example 3.1. A CW radar (MTI) is operating at a PRF of 1 kHz, find the lowest blind speed. If it is operating at 2 cm wave lengths.

Solution. Lowest blind speed $= n.\frac{\lambda}{2} \times \text{PRF}$

We have, PRF $=$ 1 kHz $=$ 1000 Hz

$\lambda = 2 \text{ cm} = 0.02 \text{ m}$

for the calculation of lowest blind speed, the value of $n = 1$,

By puttting the values in above equation, we get

$$\text{Lowest blind speed} = 1 \times \frac{0.02}{2} \times 1000 = 10 \text{ m/s}$$

or

$$= \frac{10 \times 60 \times 60}{1000} = 36 \text{ km / hr}$$

$$\boxed{\text{Lowest Blind Speed} = 36 \text{ km/hr}}$$ **Ans.**

Example 3.2. If two MTI radar system are operating at the same PRF but with the different operating frequencies. And the 2nd blind speed of one is equal to fourth speed of the other than find the ratio of their operating frequencies.

Solution. Let the operating frequency of first radar $= f_a$

Operating frequency of second radar $= f_b$

Now, second blind speed of first radar $= \left[2 \times \frac{C}{2f_a}\right] \times PRF$

and, fourth blind speed of second radar $= \left[4 \times \frac{C}{2f_b}\right] \times PRF$

$$\text{Therefore,} \quad \left[2\times\frac{C}{2f_a}\right]\times PRF = \left[4\times\frac{C}{2f_b}\right]\times PRF$$

or
$$\frac{2}{f_a} = \frac{4}{f_b}$$

or,
$$\frac{f_a}{f_b} = \frac{2}{4}=\frac{1}{2}$$

Their operating frequency ratio will be,

$$\boxed{\frac{f_a}{f_b} = \frac{1}{2}}$$ **Ans.**

Example 3.3. A pulse doppler radar has a carrier frequency of 9 GHz and PRF of 400 Hz. Find its blind doppler frequencies and the radial velocity of target which would be undetected by the radar.

Solution. Doppler frequencies (f_d) = $n \,.\, PRF$

We have, $PRF = 4000$ Hz

Than Doppler frequencies $= n(4000)$

$= 1 \times 4000 = 4000$ Hz

$= 2 \times 4000 = 800$ Hz

$= n \times 4000 = n \,.\, 4000$ Hz.

What radial target velocities would be undetected by the radar ?

We have, operating frequency = 9 GHz = 9×10^9 Hz

$$\lambda = \frac{c}{f}=\frac{3\times10^8}{9\times10^9}=3.33\,\text{cm}$$

$$\lambda = 0.03 \text{ m.}$$

Now, we know

$$V_r \text{ (blind speed)} = \frac{n\,.\,PRF\,.\,\lambda}{2} = \frac{n\times(4000)\times(0.03)}{2}$$

$$\boxed{V_r = n\,.\,66.67\text{ m/s}}$$ **Ans.**

Any integer multiple of 66.67 m/s is undetected by this radar system.

Example 3.4. What modification can be made to the radar system (used in problem 3) that would eliminate blind speeds below 2000 m/s and find out the maximum unambigious range of that modified radar system ?

Solution. We have $V_r = 2000$ m/s

We know that
$$V_r = \frac{n\,.\,PRF\times\lambda}{2} \qquad \text{(since } n = 1\text{)}$$

By putting the values,
$$2000 = \frac{1\times PRF\times 0.033}{2}$$

$$PRF = \frac{2\times2000}{0.033}=\frac{4000}{0.033}=121212\,\text{Hz}$$

$$\boxed{PRF = 121212\,\text{Hz}=121.21\,\text{kHz}}$$ **Ans.**

Maximum unambiguous range is given by formula,

$$R_{\text{unam}} = \frac{C\,.\,\Delta t}{2}$$

where $\Delta t = PRP = \frac{1}{PRF}$

$$\Delta t = \frac{1}{121212} = \frac{1}{121.21 \times 10^3}$$

By putting the values, we get

$$R_{\text{unam}} = \frac{3 \times 10^8}{2} \times \frac{1}{121.2 \times 10^3}$$

$$= \frac{3 \times 10^8 \times 10^{-3}}{2 \times 121.21} = \frac{3 \times 10^5}{242.42}$$

$$= 0.0123752 \times 10^5 \text{ m}$$

$$\boxed{R_{\text{unam}} = 1237.52 \text{ m}}$$

or $\boxed{R_{\text{unaml}} = 1.237 \text{ km.}}$ **Ans.**

Example 3.5. The MTI radar is used by a traffic control police to measure the speed of vehicles. If the dopper frequency shift measured from the moving vehicle is 1.5 kHz. Calculate the speed of vehicle, if radar is operating at 1 GHz with PRF of 1000 Hz.

Solution. The speed of the vehicle = $V_r = \frac{f_d \times \lambda}{2}$ (where f_d = dopper frequency)

$$V_r = \frac{1500 \times 0.3}{2} \text{ m/s} \qquad \left(\text{since } \lambda = \frac{C}{f} = \frac{3 \times 10^8}{1 \times 10^9} = 0.3 \text{ m}\right)$$

$$V_r = 225 \text{ m/s}$$

$$V_r = \frac{225 \times 60 \times 60}{1000} \text{ km/hr}$$

$$\boxed{V_r = 810 \text{ km/hr}}$$ **Ans.**

IMPORTANT RELATIONS

1. $V = \frac{V_1}{2} - \frac{V_2}{2} = K \sin \pi f_d T \cos \left\{2\pi f_d \left(t - \frac{T}{2}\right) - \phi_0\right\}$

   (The output from the subtractor)

2. $f(t) - 2f(t+T) + f(t+2T)$ (The output of the adder)

3. $f(t) - f(t+T) - f(t+T) + f(t+2T)$ (The output of double delay canceller)

4. $f_d = \frac{n}{T} = nf_r$

5. $V_n = n\frac{\lambda}{2}T = \frac{n\lambda f_r}{2}$

6. $V_n = \frac{n\lambda f}{1.02} = n\lambda f_r$

7. $f_d = 2\left(\frac{v}{\lambda}\right)\cos\theta$

8. $\Delta f_d = \frac{2v}{\lambda}\sin\theta\,\Delta\theta$

(7 and 8: Moving platform MTI)

9. $2A \sin\left(\frac{\Delta\phi}{2}\right) = A\Delta\phi$

10. $W(f) = |G(f)|^2 = |g_0|^2 \exp\left[-a\left(\frac{f}{f_0}\right)^2\right]$

11. $t_0 = \frac{n_B}{f_r} = \frac{\theta_B}{\theta_S}$

where, V = Video signal output from the subtractor
K = Amplitude of video signal
ϕ_0 = Phase shift
V_n = nth blind speed
f_d = Doppler frequency shift
v = Platform speed
θ = Azimuth angle
$W(f)$ = Clutter power spectrum as a function of frequency
$G(f)$ = Fourier Transform of the input waveform
f_0 = Radar carrier frequency
a = Parameter which depends on the clutter
t_0 = Finite time
nB = Number of hits received
f_r = PRF
θ_B = Antenna beamwidth
θ_S = Antenna scanning rate

SUMMARY

- **Principle of MTI Radar.** When an aircraft echo and a clutter echo appear in the same radar resolution cell, the aircraft might not be detected. But the doppler effect permits the radar to distinguish moving targets in the presence of fixed target even the echo signal from fixed targets has comparatively greater magnitude than the moving targets such as aircraft.
- There is a type of MTI, which does use the information directly from the doppler but it observing the area map scan to scan. Only moving objects, which are, changes their position scan to scan from mapping operation are displayed. This type of MTI is known as **Area MTI**.
- The delay line must introduce a delay equal to the pulse–repetition interval.
- The moving target detector (MTD) employs all the techniques to achieve a good overall moving target detection performance. The MTD is an enhanced configuration of moving target indicator (MTI) that combines a series of features to improve clutter rejection and target detection. The MTD term is applied to the Doppler processing system used in airport surveillance radars.
- This processor has been integrated into the ARTS-III system at the National Aviation Facilities Experimental Center, Atlantic City, New Jersey (NAFEC) and has undergone careful testing during the summer of 1975.

UNSOLVED PROBLEMS

1. A CW radar is operating at a PRF of 10 kHz and be operating frequency is 2 GHz, find the lowest blind speed of the MTI radar.
2. Two MTI radar system are operating at the same PRF but have different operating frequencies. If the third blind speed of one is equal to sixth blind speed of the other, find the ratio of their operating frequencies.
3. An MTI radar in operating at 1 GHz will a PRF of 1 kHz. Calculate the first three blind speeds.
4. Calculate the two lowest blind speeds for a MTI radar operating at λ = 10 cm and PRF = 1000 Hz.
5. If a MTI radar is used by a traffic control police to measure the speed of car. If the Doppler frequency shift measurement from the car in 1000 Hz. Calculate the speed of car. The PRF of radar in 1.5 kHz and operating at frequency 10 GHz.
6. A pulse Doppler radar has a carrier frequency of 10 GHz and PRF of 5000 Hz.
   (*a*) What are its blind doppler frequencies ?
   (*b*) What radial target velocities would be undetected by the radar ?
   (*c*) What modification can be made to the radar that would eliminate blind speeds below 2000 Knots (use PRF).
   (**Hint**. 2000 Knots = 10^{16} m/s)
   (*d*) What would the maximum unambiguous range of the modified radar be ?

REVIEW QUESTIONS

1. What do you understand by term MTI ?
2. Describe the working of Non-coherent MTI radar with the help of block diagram.
3. What are the advantages of digital MTI over analog MTI ?
4. Explain MTI radar functional block diagram using Power Amplifier.
5. Discuss the limitations of MTI radar.
6. What is blind speed and how you can avoid it ?
7. What is the necessity of delay line canceller ? Describe various types of delay-lines used in MTI radar.
8. Differentiate between MTI and Pulse Doppler radar.
9. Explain the working of coherent AMTI radar with the help of block diagram.
10. Describe coherent MTI radar with neat diagram. **[UPTU, 2003]**
11. Explain the MTI radar block diagram.
12. Discuss various aspect of MTI from a moving platform.
13. What is subclutter visibility ?
14. Explain the principle and working of MTI radar.
15. Give a brief description of MTI radar. **[UPTU 2004]**
16. Draw a block diagram and explain the operations of an MTI Radar. What are delay lines in such radar ?
17. What is Non-coherent MTI ?
18. Explain the use of delay line cancellers. Give its filter characteristics. **[UPTU 2006]**
19. Explain MTI radar and give its limitations. **[UPTU, 2006]**

20. Discuss the limitations to MTI performance.

21. Discuss the pulse Doppler radar.

22. Write short notes on :

(*a*) Staggered PRF

(*b*) Delay lines and cancellers

(*c*) Doppler range gate filters

(*d*) Digital MTI signal processing

(*e*) Blind phase

(*f*) Blind speed

23. What are the various aspects of MTI from a moving platform ?

24. Explain MTD with help of simple block diagram.

❑❑❑

CHAPTER

4

TRACKING RADAR

4.1. INTRODUCTION

INSIDE THIS CHAPTER

In this chapter students will be studying about the Tracking of the targets with the help of Tracking Radar. Tracking means tracking the path of the targets by means of the radar. The following characteristics of the targets are important:

1. Range of the target
2. Velocity of the target
3. Azimuth angle of the target

Different types of techniques of tracking the targets (such as lobe switching, conical switching, monopulse tracking) are being explained in this chapter. A special radar, Synthetic Aperture Radar (SAR) is also being discussed in this chapter.

4.2. TRACKING WITH RADAR

A tracking radar system measures the different co-ordinate of a target such as range, elevation angle, azimuth angle, and doppler shift frequency to determine the target path and to predict its future position. All or any of the coordinates may be used to predict the future position. In general it is the method by which angle tracking is accomplished may be called the tracking radar. One of the most useful features of radar is the ability of a radar set to continuously predict the next location of its target from the information being received from the target and to align itself to continuously point at that predicted location. When this is occurring, the radar set is said to be **tracking the target.**

4.2.1. Radar Servo Tracking System

One of the most basic tracking system designs is the servo tracking system, as shown in the Fig. 4.1. Here, the radar antenna is initially trained on a target after which it automatically remains pointed at the target as it follows its motion. Furthermore, the system provides continuous position information to the operator and possibly to a fire control system. The antenna is rotated by a motor which provides a negative position feedback signal to a controller. This subsystem is known as a **servo-mechanism.**

The range and doppler frequency shift can also be continuously tracked, if desired, by a servo-control-loop actuated by an error signal generated in the radar receives. In the continuous radar a servomechanism actuated by an error signal make the antenna beam in angle. There are various methods for generating the error signal may be classified as given :

1. A sequential lobbing
2. A conical scan
3. A mono pulse.

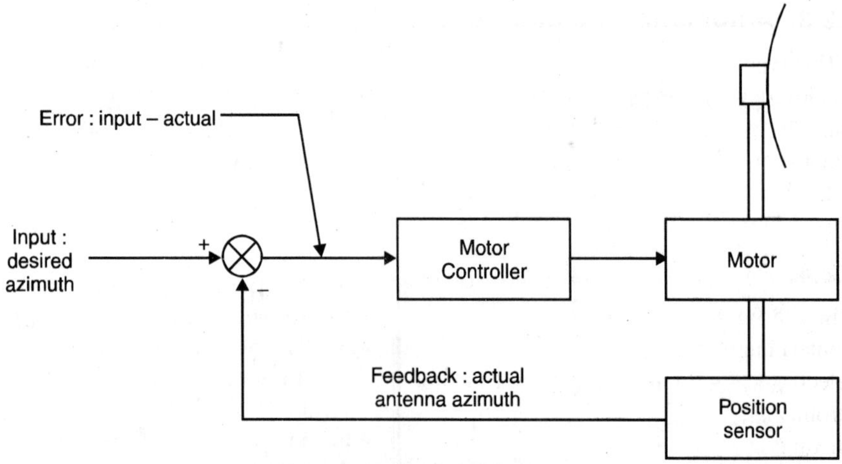

Fig. 4.1. Servo Mechanism.

Tracking radars information may be given to the display unit or may be supplied to an automatic computer, which determines the target path and predicts its probable future course. Tracking radar must find its target before it can track. There are some radar, which works as search radar to find the target before switching to a tracking mode. Due to some operational limitation, single radar for both the search and the tracking mode is not favourable. If the antenna pattern is a narrow pencil beam and if the reach area is large, a relatively long time might be required to find the target. Therefore, separate search radar is employed to provide the information related to the position of the target to the tracking system. When separate search radar employed for giving the information of the target to the tracking radar system, is called acquisition radar.

4.2.2. Types of Tracking Radar

There are many types of tracking radar, which are being used to track the targets. They are as follows :

1. Single Target Tracker (STT)
2. Track While Scan (TWS)
3. Automatic Detection And Track (ADT)
4. Phased Array Tracking

4.2.2.1. Single Target Tracker (STT)

A single target tracker is used where continuous tracking of a single target with the higher data rate is required. Generally this type of tracking is used in controlling the missile movements as it is carried out by a Missile Guidance Radar (MGR).

In this type of tracker a closed loop servomechanism is used to keep the angle-coordinates error very small. If the errors are less than there will be more accuracy.

4.2.2.2. Track While Scan (TWS)

TWS radar scans their beam over relatively large areas. It rapidly scans in the angular sector to keep track of the targets, may be move one target at a time. The radar computer still measures returned power as a function of beam location to provide tracking but the large scanning area enables the radar to still see the target even if track has beam broken or lost. However, this large scan makes the TWS highly vulnerable to ECM jamming. An illustration of TWS radar is shown in Fig. 4.2.

4.2.2.3. Automatic Detection and Track

This type of technique used in the most of the air traffic control radar. This technique is employed with the air surveillance radar, which has 7.5 to 15 RPM (Revolution per minute). It does provide updating data according to the rotation of the antenna. It can track enormous number of targets simultaneously may be few hundreds. In this antenna position is not controlled by closed loop, it works in the open loop system. It is being further discussed in chapter 10 (detection of signal in noise).

4.2.2.4. Phased Array Radar Tracking

In this technique multiple targets are tracked in the time-sharing basis by the computer. Large numbers of tracks can be scanned rapidly in this technique because of the electronically steered phased array method is used here. The antenna beam is scanned electronically make to switched on from one angular direction to another in a very small time such as few microseconds.

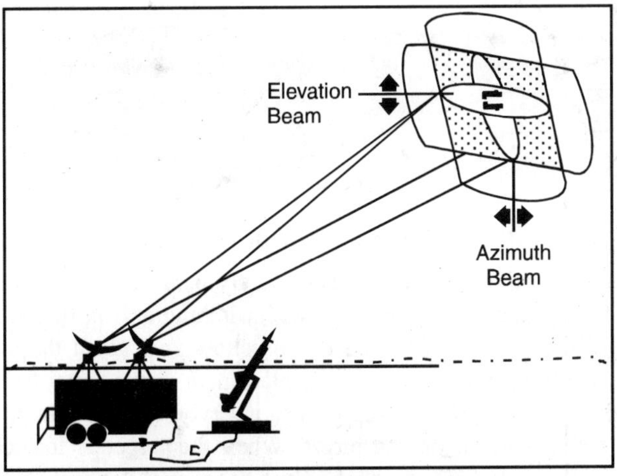

Fig. 4.2. Track-While-Scan Radar.

4.3. BLOCK DIAGRAM OF TRACKING RADAR

Figure 4.3 shows the typical block diagram of Tracking Radar. Most tracking radar use angular information as the basis for tracking operation. For better accuracy it is important that radar concentrates on one target at time. Range gating/Doppler filtering can be used for that purpose. Time and frequency control for range and doppler gating, is done in range and doppler trackers respectively. The angular error signal for the desired target to be tracked, are developed in the error demodulator block which is also controlled by range/doppler gate generation block and than fed back to the steerable antenna in a closed loop for tracking.

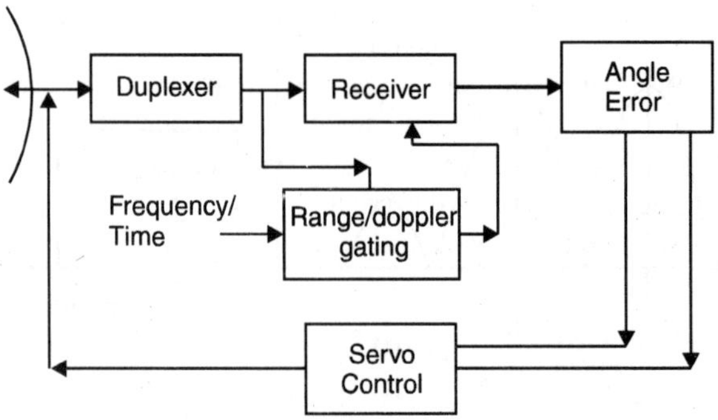

Fig. 4.3. Typical Block Diagram of Tracking Radar.

4.4. SEQUENTIAL LOBBING

A single beam is switched between two sequential angular positions to obtain on angle measurement. This is called sequential lobbing, lobe switching or sequential switching. Here the direction of the antenna beam is rapidly switched between two position *A* & *B* as shown in Fig. 4.4. The echo signal from the target will fluctuate at

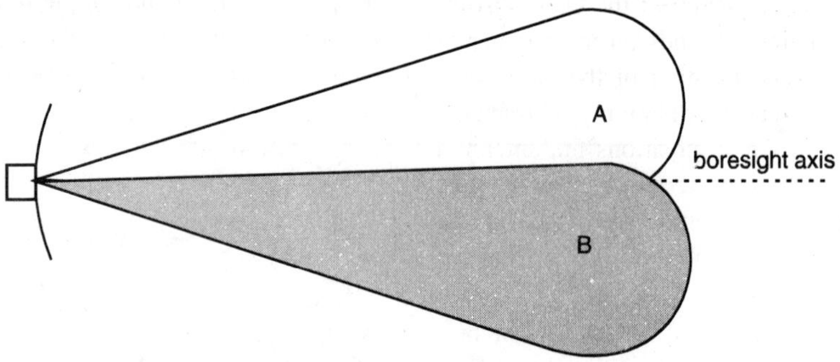

Fig. 4.4. Dual Beam Polar Pattern in Sequential Lobbing.

the switching rate, unless the target reaches exactly between two directions. The error signal obtained from a target not located on the switching axis is shown in Fig. 4.5. The difference between the amplitude of these two echo signal given the error signal. The error signal may be defined as the angular displacement of the target from the switching axis. These tracking error signals are applied to a servomechanism unit, which attempts to position the antenna beam on the target. When the target is located at references direction than the angular error is zero. Thus sequential lobbing with two position switched beam is used for tracking a target in only one place. The sign of the difference determine the direction of the antenna to move in order to align the switching axis with direction of the target.

Four switching position are required to obtain the angle measurement in the orthogonal coordinate. Thus a two-dimensional sequentially lobbing radar consists of a cluster of four feed horns illuminating a single reflector antenna, arranged so that the right-left, up-down sectors are covered by successive antenna position. A cluster of five feed horns might also be used, with a control feed used for transmission and four outer feeds used for reception on a sequential basis.

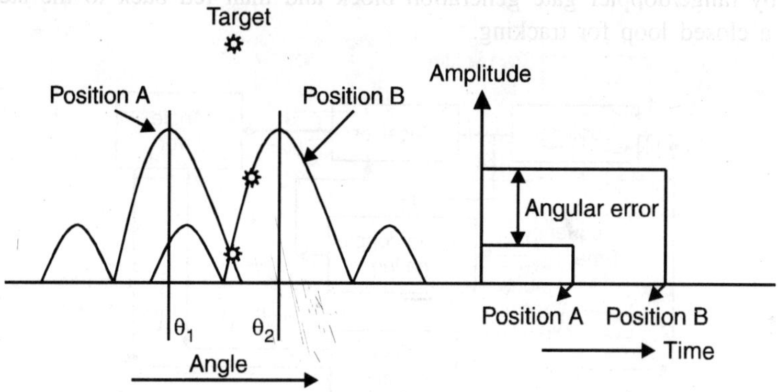

Fig. 4.5. Rectangular Representation and Error Signal.

4.5. CONICAL SCANNING

In the conical scanning, the squinted beam is scanned rapidly and continuously in a circular path around the axis. The angle between the axis of rotation and the axis of the antenna beam is called the squint angle. If a target is present within squint angle, then the echo signal from the target will be amplitude modulated at a frequency equal to the rotation frequency of the beam also called conical scan frequency. The amplitude of the modulation depends on the angular distance between the target direction and the rotation axis. The location of the target in two angle co-ordinate determines the phase of the conical-scan modulation is extracted from the echo signal and applied to a servo control system which positions the antenna on the target both in azimuth and elevation Two servo motors are required one for azimuth and the other for elevation. The conical scan modulation becomes zero and thus the target is tracked accurately by conical scanning, both in azimuth and elevation.

For true tracking, the scan frequency must be as per the Nyquist criteria for the sampling rate. In pulsed radar it must be at least are four pulses per scan, two for generating azimuth error signal and two for generating elevation error signal. The actual seen frequency also depends upon the scan mechanism. The antenna beam is squinted and scanned either mechanically by offsetting the feed and rotating it or electronically with the help of phase shifters. Electronic scan is very fast than the mechanically scan. Typical scan rates 30- to 40-scan/sec.

Block diagram. A B/D of the conical scan tracking radar system is as shown in Fig. 4.6 The antenna is mounted such that it can be positioned both in elevation with the help of respective motors. The scan antenna beam generates two signals, the for azimuth detector and the other for elevation detector, which are 90 degree out of phase.

The receiver is a conventional superhetrodyne except for features related to the conical scan tracking radar. The error signal is extract in the video after the second detector. The error signal is compared with the elevation and azimuth references signal, in the angle detectors, which are phase sensitive detection. Phase sensitive detector is a non linear device in which the input signal is mixed with a reference signal. This angle detection produces a D.C. voltage, which is proportional to the error, and the sign is an indication of the direction of the error. The angle error detector output is amplified and drive the antenna elevation and azimuth servomotors. The angular position of the target may be determined from the elevation and azimuth of the antenna axis.

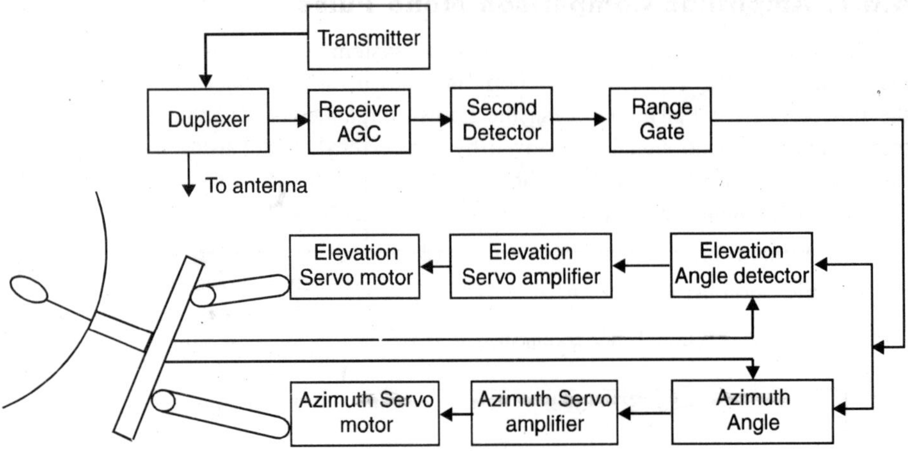

Fig. 4.6. Block Diagram of Conical Scan System.

Conical Scan systems require a minimum amount of hardware and therefore are commonly used on inexpensive, mobile systems such as AAA or mobile SAM sites. They suffer the serious disadvantage of not being able to see a target outside their narrow scan patterns. This means that not only is a second radar required to help it find the target (to "acquire the target") but also the tracked aircraft can easily "escape" if it is successful in breaking track since the Conical Scan radar cannot see the target, except in the track mode.

4.6. MONO PULSE TRACKING

There are two disadvantages connected with the use of lobe scanning and conical scanning which are as follows :

1. The motion of the antenna is more complex in lobe switching and conical scanning.
2. In conical scan, a minimum of four pulses is required. The difficulty here is that if the targets cross-section, the tracking accuracy may be degraded.

The effect of the fluctuating echo can be sufficiently serious in some application severely limits the accuracy of those tracking radar which requires many pulses to be processed in extracting the error signal.

The above problems can be overcome by using the single pulse only. There are several methods by which angle error information might be obtained with a single pulse. The tracking techniques which derive angle error information on the basis of single pulse, is known as mono pulse tracking or simultaneous lobbing more than one antenna beam is used simultaneously in there method where as in conical scan or lob switching tracker, which uses one antenna beam on a time-shared basis. There are several methods by which a mono pulse angle measurement can be made. The most popular method is the amplitude–comparison mono pulse, which compares the amplitude of the signals simultaneously received in multiple squinted beams to determine the angle.

4.6.1. Amplitude Comparison Mono Pulse

In an amplitude comparison mono pulse system, four feeds are used with one paraboloid reflector. There are four horn antennas displaced about the central focus of the reflector. The Transmitter feeds the horns simultaneously so that a sum signal is transmitted. The echo signal is received by a receiver duplexer using a hybrid ring (or rat-race) to provide the following three signals :

1. Sum signals $(A + B + C + D)$
2. Azimuth error signal $= (A + C) - (B + D)$
3. Elevation error signal $= (A + B) - (C + D)$

A diagram of Monopulse Radar Beam Patterns is shown in Figure 4.7.

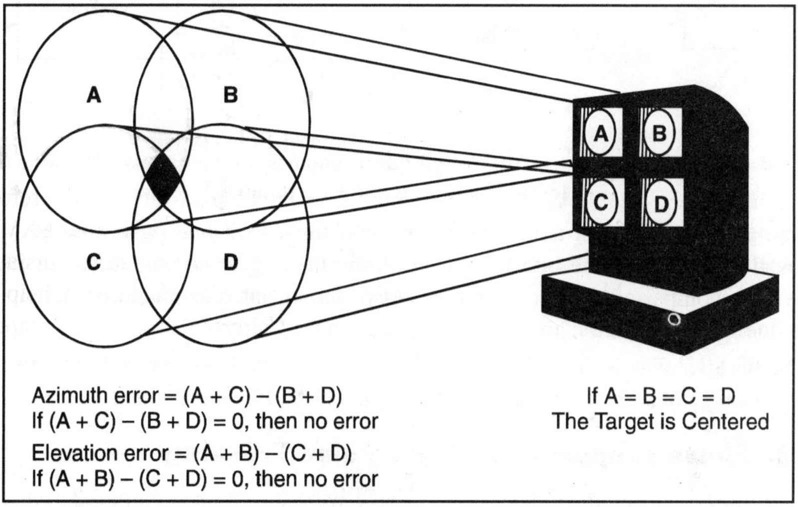

Fig. 4.7. Diagram of Monopulse Radar Beam Patterns.

In an amplitude comparison mono pulse tracking technique, it is important the signal arriving at various feeds are in phase. It is not a problem in reflector antenna with feeds that are physically small. In case of arrays where the antenna surface is very large signals arriving from different off-axis angles present different phases to different segment into which the array has been divided. Their phases need to be equalized before error signals are developed. The sum pattern is used for transmission, while both the sum pattern and the difference pattern are used in reception. The signal received with the difference pattern provides the range measurement and is also used as references to extract the sign of the error signal. Signals received from the sum and difference patterns are amplified separately and combined in a phase sensitive detector to produce the error signal characteristics.

A block diagram of the amplitude comparisons monopulse tracking radar for a single angular coordinates shown in Fig. 4.8.

The receiver has three separate input channels consisting of three mixers, common local oscillator, three IF amplifiers and three detector. The elevation and azimuth error signals are used to drive a servo amplifier and a motor in order to position the antenna in the direction of target. The output of sum channel is used to provide the data generally obtained from a radar receiver so that it can be used for application like automatic control of the firing weapon.

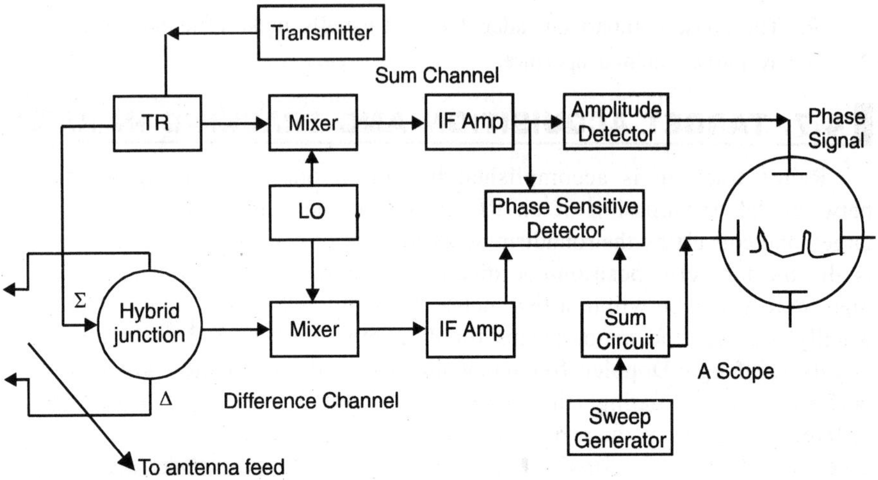

Fig. 4.8. A Block Diagram of the Amplitude Comparisons Monopulse Tracking Radar for a Single Angular Coordinates.

The advantage of this technique is that it obtains with one pulse (unlike 4 pulses in conical scanning) all the information regarding the target and able to locate the target in less time comparing to other methods. Also, mono pulse technique is not subject to errors due to the variation in target cross-section.

The disadvantage is that it requires two extra R_x channels and more complex duplexer feeding arrangement, which makes the system bulky and more expensive.

4.6.2. Phase Comparison Mono Pulse Tracking

The tracking techniques discussed till now were based on a amplitude comparison of the echo signals received from two or more antenna position. The angular error was generated by the difference in amplitude in the several antenna position. The angular error may also be determined by comparing the difference in the phases between the signals received from the two separate antennas. The antennas used in phase comparison monopulse tracking are not offset from the axis as used in amplitude comparison trackers. It is essential that the amplitude of the target echo signal is the same from the each antenna, but the phases are different.

Tracking radar, which operates with phase information, might be called **simultaneous phase comparison radar**, or phase comparison mono pulse. The measurement of angle of arrival by comparison of the phase relationship in the signals from the separated antennas of a radio interfero meter has been widely used by the radio astronomers for precise measurement of the position of radio stars. The interferometer as used by the radio astronomer is a passive instrument the source of energy being radiated by the target itself.

Tracking radar, which operates with phase information, is similar to an active interferometer.

An additional antenna and receiving channel is necessary in order to track in two orthogonal co-ordinates. In the phase comparison radar, four antennas are required, one of these antenna for transmitter only, while the other three for receiver.

This technique has not been as widely used as the other techniques discussed earlier. There are two main drawback why this might be so.

1. The side lobes levels, which result higher than the single antenna.
2. The phase comparison radar does not usually make efficiently use of the total available antenna aperture.

4.7. TARGET ACQUISITION AND TRACKING IN RANGE

Range tracking is accomplished by continuously measuring the time delay between the transmission of an RF pulse and the echo signal returned from the target, and converting the roundtrip delay to units of distance. The range measurement is the most precise position-coordinate measurement of the radar; typically, with high SNR, it can be within a few meters at hundreds-of-miles range. Range tracking usually provides the major means for discriminating the desired target from other targets (although Doppler frequency and angle discrimination are also used) by performing range gating (time gating) to eliminate the echo of other targets at different ranges from the error-detector outputs. The range-tracking circuitry is also used for acquiring a desired target. Range tracking requires not only that the time of travel of the pulse to and from the target be measured but also that the return is identified as a target rather than noise and a range-time history of the target be maintained.

Although this discussion is for typical pulse-type tracking radars, range measurement may also be performed with CW radars using FM-CW, a frequency-modulated CW that is typically a linear-ramp FM. The target range is determined by the range-related frequency difference between the echo-frequency ramp and the frequency of the ramp being transmitted. The performance of FM-CW systems, with consideration of the Doppler Effect.

4.7.1. Target Acquisition

The acquisition radars are those radar which are being employed to search the target to be tracked and then designate the target position to the tracker. There radar have large search large an surveillance radar. Normally a tracker radar uses a narrow beam (pencil) to track the target in angle accurately. So it is difficult for the tracker to search the target in large area due narrow beam width. The acquisition of the target for the purpose of tracking is performed by the acquisition radar, and in known as the acquisition of the target.

There are various method of scanning pattern used with narrow beam (pencil) antennas to find a target.

1. Helical scanning beam
2. Palmer scan
3. Spiral scan
4. Raster or TV scan
5. Nodding scan

1. Helical Scan Beam

In this method antenna is scanning in azimuth continuously and at the same time it is raising or lowering in elevation scanning. It makes a helix track in the space. If fire control radar used in the search made than the antenna will rotate at certain RPM and covered some elevation angle.

2. Palmer Scan

This type of scanning is used where a search area is larger in one dimension than another. Palmer scan consists of a conical scan around the axis of rotation.

The palmer scan reduces to the conical scan where the axis of rotation is stationary. To generate palmer scan pattern, same mechanism of conical scan may be employed.

3. Spiral Scan

It covers an angular reach area with the circular symmetry. In the spiral scan volume do not receive the same signal unless the speed of the scanning is changed during the complete scan cycle.

4. Raster or TV Scan

The raster scan reach the limited area in the rectangular in the shape. In this scan the reach area is painted in uniform manner unlike the spiral and palmer scan.

5. Nodding Scan

The nodding scan is produced by oscillating the antenna beam slowly in azimuth and rapidly in elevations. Generally this method is used in height finding radars. It covers the hemispherical area with the narrow beam width, that is, in elevation it is up to 90° and in azimuths up to 360°.

The first function of the range tracker is acquisition of a desired target. Although this is not a tracking operation, it is a necessary first step before range tracking or angle tracking may take place in typical radar. Some knowledge of target angular location is necessary for pencil-beam tracking radars to point their typically narrow antenna beams in the direction of the target. This information, called designation data, may be provided by surveillance radar or some other source. It may be sufficiently accurate to place the pencil beam on the target, or it may require the tracker to scan a larger region of uncertainty. The range-tracking portion of the radar has the advantage of seeing all targets within the beam from close range out to the maximum range of the radar. It typically breaks this range into small increments, each of which may be simultaneously examined for the presence of a target. When beam scanning is necessary, the range tracker examines the increments simultaneously for short periods, such as 0.1 s, makes its decision about the presence of a target, and allows the beam to move to a new location if no target is present. This process is typically continuous for mechanical-type trackers that move the beam slowly enough that a target will remain well within the beam for the short examination period of the range increments.

Target acquisition involves consideration of the S/N threshold and integration time needed to accomplish a given probability of detection with a given false-alarm rate similar to surveillance radar. However, high false-alarm rates, as compared with values used for surveillance radars, are used because the operator knows that the target is present, and operator fatigue from false alarms when waiting for a target is not involved. Optimum false-alarm rates are selected on the basis of performance of electronic circuits that observe each range interval to determine which interval has the target echo.

A typical technique is to set a voltage threshold sufficiently high to prevent most noise peaks from crossing the threshold but sufficiently low that a weak signal may cross. An observation is made after each transmitter pulse as to whether, in the range interval being examined, the threshold has been crossed. The integration time allows the radar to make this observation several times before deciding if there is a target present.

The major difference between noise and a target echo is that noise spikes exceeding the threshold are random, but if a target is present, the threshold crossings are more regular. One typical system simply counts the number of threshold crossings over the integration period, and if crossings occur for more than half the number of times that the radar has transmitted, a target is indicated as being present. If the radar pulse repetition frequency is 300 Hz and the integration time is 0.1 s, the radar will observe 30 threshold crossings if there is a strong and steady target. However, because the echo from a weak target combined with noise may not always cross the threshold, a limit may be set, such as 15 crossings, that must occur during the integration period for a decision that a target is present. For example, performance on a non-scintillating target has a 90% probability of detection at a 2.5 dB-per-pulse SNR and a false alarm probability of 10–5. The AN/FPS-16 and AN/FPQ-6 instrumentation radars use these detection parameters with 10 contiguous range gates of 1000 yd each for acquisition. The 10 gates give coverage of a 5-nmi range interval at the range where the target is expected, possibly from coarse range designation from search radar.

4.7.2. Range Tracking

Once a target is acquired in range, it is desirable to follow the target in the range coordinate to provide distance information or slant range to the target. Appropriate timing pulses provide range gating so the angle-tracking circuits and AGC circuits look at only the short range interval, or time interval, when the desired echo pulse is expected. The range tracking operation is performed by closed-loop tracking similar to the angle tracker. Error in centering the range gate on the target echo pulse is sensed, error voltages are generated, and circuitry is provided to respond to the error voltage by causing the gate to move in a direction to re-center on the target echo pulse.

The range-tracking error may be sensed in many ways. The most commonly used method is the early- and late-gate technique (see Fig. 4.9). These gates are timed so that the early gate opens at the beginning of the main range gate and closes at the center of the main gate. The late gate opens at the center and closes at the end of the main range gate. The early and late gates each allow the target video to charge capacitors during the time when the gates are open. The capacitors act as integrators. The early-gate capacitor charges to a voltage proportional to the area of the first half of the target video pulse, and the late-gate capacitor charges negatively proportionally to the late half of the target video. When the gates are properly centered about a symmetrical video pulse, the capacitors are equally charged. Summing their charge voltages yields a zero output.

When the gates are not centered about the target video, so that the early gate extends past the center of the target video, the early-gate capacitor charged positively receives a greater charge. The late-gate sees only a small portion of the pulse, resulting in a smaller negative charge. Summing of the capacitor voltages results in a negative output. Over a range of errors of approximately ±1/4 of the target-video pulse width, the voltage output is essentially a linear function of timing error and of a polarity corresponding to the direction of error. During acquisition, the target is centered in the 1000 yd acquisition gate by range-tracking techniques described as follows, and the gate is reduced to approximately the width of the radar transmit pulse for normal tracking.

Many radar range-tracking systems use high speed sampling circuitry to take three to five samples in the vicinity of the echo video pulse. The amplitudes of the samples on the leading and lagging halves of the pulse are compared for range-error sensing similar to the comparison of amplitudes in the early-late-gates range tracker.

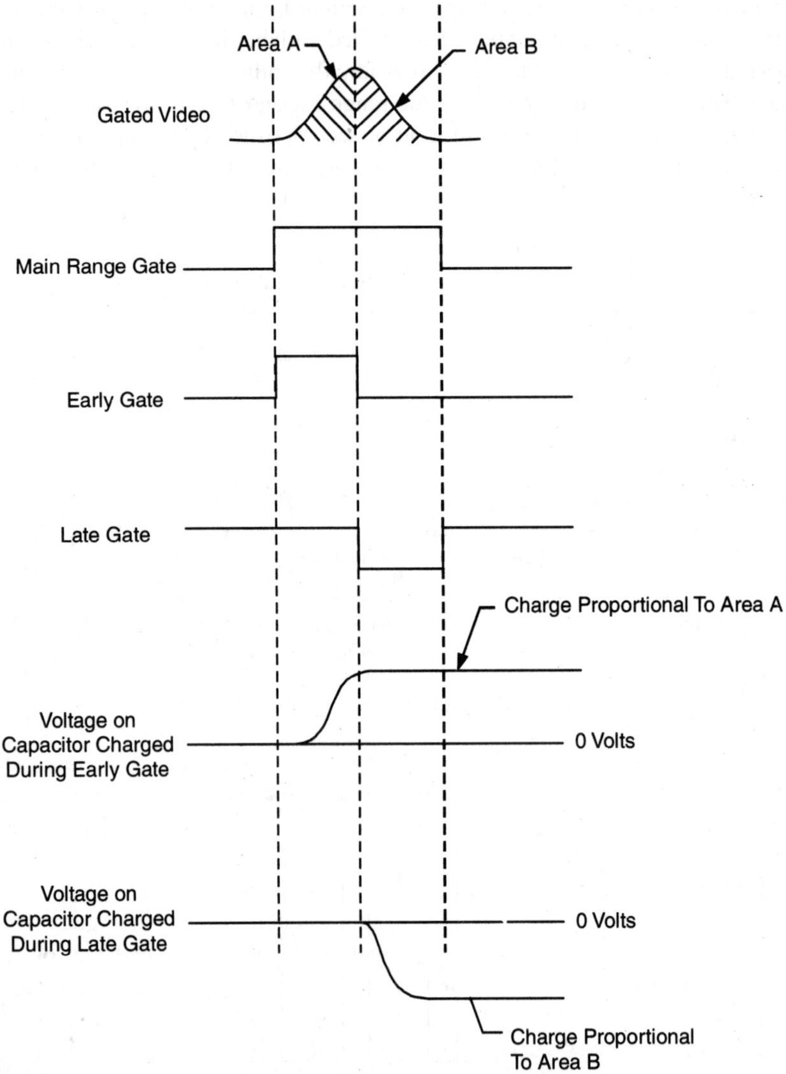

Fig. 4.9. Early- and late-gate range-error-sensing circuit.

In some cases, leading- or lagging-edge range tracking is desired. This has been accomplished in some applications by simply adding a bias to move the error-sensing gates either to lead or lag the center of the target. This provides some rejection by the gates of undesired returns that might occur near the target, such as the echoes from other nearby targets. Threshold devices are also used as leading- or lagging-edge trackers by observing when the target video exceeds a given threshold level. The point of crossing the threshold is used to trigger gating circuits to read out a target range from timing devices or to generate a synthetic target pulse.

The range-tracking loop is closed by using the range-error-detector output to reposition range gates and correct range readout. One technique uses a high-speed digital counter driven by a stable oscillator. The counter is reset to zero at the time of the transmit pulse. Target range is represented by a number stored in a digital register, as shown in Fig. 4.10. A coincidence circuit senses when the digital counter reaches the number in the range register and generates the range gate, as indicated in the block diagram shown in Fig. 4.11. A range error sensed by the range error detector results in an error voltage that drives a voltage-controlled variable-frequency oscillator to increase or decrease the count in the range register, depending on the polarity of the error voltage. This changes the number in the range register toward the value corresponding to the range of the target. Range readout is accomplished by reading the number in the register, where, for example, each bit may correspond to a 2-yd range step.

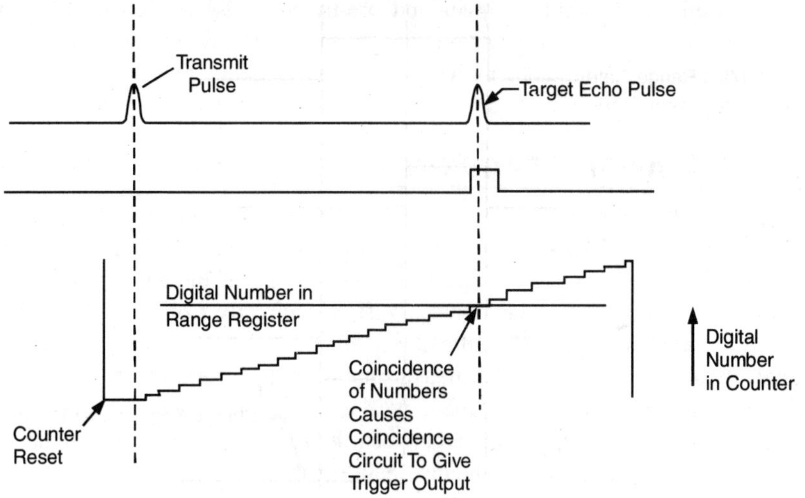

Fig. 4.10. Digital range tracker operation.

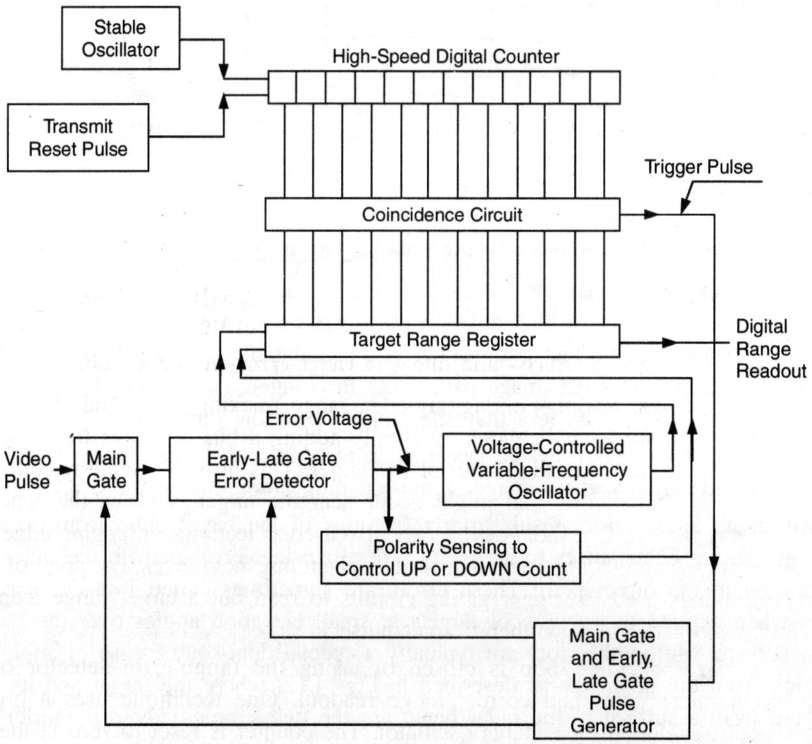

Fig. 4.11. Block diagram of a digital range tracker.

Another technique is to use a pair of oscillators—one controlling the transmitter trigger and the other controlling the range gate. The range rate is controlled by the beat frequency between the oscillators, where one is frequency-controlled by the range error-detector-output voltage. The beat frequency is a small fraction of one Hz and is better visualized as a phase rate between the transmit pulse cycle and cycle of the range gate. The changing phase causes the range gate to follow a moving target.

The electronic range tracker is inertia less, allowing any desired slew speed, and provides flexibility for conveniently generating acquisition gates for automatic-detection circuitry as well as transmitter trigger and pre-trigger pulses. Tracking bandwidth is usually limited to that necessary for tracking to minimize loss of track to false targets and countermeasures. Many other electronic range-tracking techniques also offering most of these advantages are used.

4.8. LOW ANGLE TRACKING

Conventional monopulse radar tracking equipment suffers severe degradation of tracking accuracy when tracking targets at low angles. Figure 4.12 shows a radar that tracks a target at a low elevation angle, near the surface of the earth, can receive two echo signals from the target. One signal is reflected directly from target, and the other arrives via the earth's surface. The direct and the surface-reflected signals combine at the radar to determine an angle measurement that differs from the actual true measurement that would have been made with a single target in the absence of surface reflections. This is due to interference between the direct reflected radar signal and the reflected image signal transverse a bounce path off the earth or sea surface. This effect becomes more serious when the target is within a beam width of the horizon and when both signals fall within the main beam.

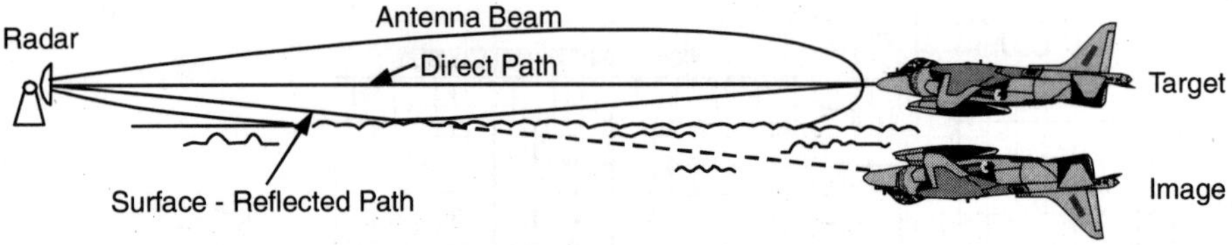

Fig. 4.12. Geometry of the radar multipath tracking condition, where the reflection from a surface appears to the radar as an image below the surface.

For example, where targets are close to the water, the direct signal and the image signal are symmetrically displaced about the horizon. A path length difference exists between the image and direct path which is a calculatable function of antenna height and elevation angle with the image path being the longer. The amplitude of the image signal therefore tends to be less than the amplitude of the direct signal.

Conventional monopulse radar systems are incapable of distinguishing between the direct reflected signal and its image signal, creating ambiguity in the display.

Multipath angle errors result from reflections of the target echo from objects or surfaces causing echo pulses to arrive by other than the direct path to the radar beam in addition to the direct path. These errors are sometimes called low-angle tracking errors when applied to tracking of targets at small elevation angles over the Earth or ocean surface. Multipath errors are typically a special dual-source case of angle noise resulting from the geometry as described in Fig. 4.12, where the target and its image reflected from a surface to the radar beam are the two sources. Over a smooth ocean surface, they are separated only in the elevation coordinate so that most of the error appears in the elevation-tracking channel. Severe elevation errors may cause some cross coupling of the error to the azimuth channel. Rough surfaces cause diffuse scattering, which can contribute errors to both azimuth and elevation tracking. Different path geometries such as non-flat land or a building may also cause a significant error to appear in the azimuth-tracking channel.

The major difficulty with low-angle tracking is that the target and its image are essentially coherent and their relative phase changes slowly and the angular error it causes is readily followed by an angle-tracking system. Furthermore, the paths are almost equal, and in most cases, they cannot be resolved by high-range-resolution techniques. Long time averages of the data do not, in practice, give target elevation; thus, the multipath angle error has no simple solution and is generally minimized by using narrow-beam antennas.

When the target is at low altitude, the multipath errors are severe, as observed in the measured data shown in Fig. 4.13. This data is the multipath error of a 2.7° beamwidth S band (3-GHz) tracking radar that is tracking an aircraft target with a beacon at 3300-ft altitude. An AN/FPS-16 tracking radar with a 1.1° beamwidth at C band (5.7 GHz) was used to simultaneously track with its narrow beam, which remained above the sea surface without significant multipath error, to provide a true target altitude reference for the data in Fig. 4.13. There is a measurement bias error (observed in Fig. 4.13) of about 0.25°.

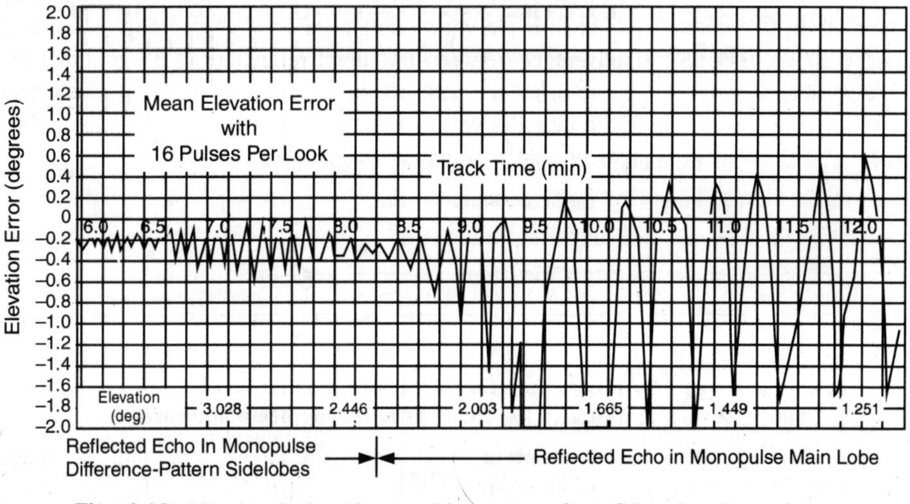

Fig. 4.13. Measured elevation-tracking error of an S-band radar using an AN/FPS-16 radar for a target elevation reference.

The tracking data from radar tracking a target with a beacon is plotted in Fig. 4.13, showing a typical multipath error illustrating the phenomenon from the region where the image enters the sidelobes to the region where it enters the main beam. There are three methods used for predicting multipath errors, depending upon where the reflected target image enters the antenna pattern. At the far range, the image enters the antenna main beam, and the error is essentially that of a two-reflector target glint error following approximately the equation,

$$e = 2h(\rho^2 + \rho \cos \varphi/1 + \rho^2 + 2\rho \cos \varphi) \quad \text{...(4.1)}$$

where e = error, same units as h, measured at the target range relative to the target

ρ = magnitude of surface reflection coefficient

h = height of target

φ = relative phase determined by geometry of direct and surface-reflected signal paths, as shown in Fig. 4.12.

Although the fluctuations in ρ and φ alter the actual tracking from the theoretical, the equation gives a good indication of the errors to be expected when tracking a point source such as a beacon. However, skin tracking of an aircraft at low elevation may result in a departure from the classic periodic error versus elevation, as illustrated in Fig. 4.13, because of an interaction between target angle scintillation and multipath error that can change the characteristics of the multipath error.

When tracking a point source target at close range, the radar main beam is above the image, but the image is seen by the difference-pattern sidelobes. The multipath errors that result are cyclic, almost sinusoidal, with an rms value predicted by the equation,

$$\sigma_E = \frac{\rho\theta_B}{\sqrt{8G_{se}\text{ (peak)}}}$$

where σ_E = rms elevation angle multipath error, same units as θ_B

θ_B = one-way antenna beamwidth

ρ = reflection coefficient.

and G_{se} (peak) is the power ratio of the tracking-antenna sum-pattern peak to the error-pattern peak sidelobe level at the angle of arrival of the image signal.

The cyclic rate may be approximated by the equation

$$f_m = \frac{hE}{2\lambda} \quad ...(4.2)$$

where f_m = frequency of cyclic multipath error, rad/s

h = height of radar antenna

λ = wavelength, same units as h

E = rate of target elevation change as seen by radar, rad/s.

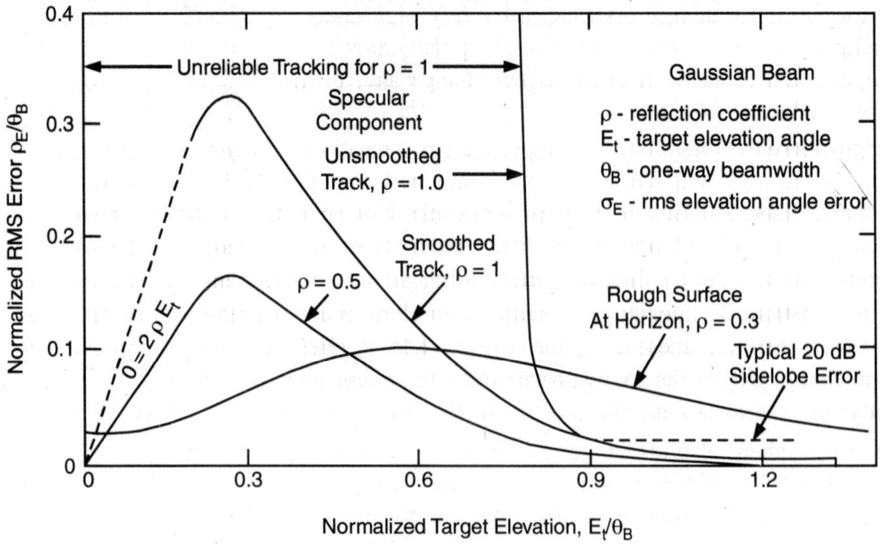

Fig. 4.14. Calculated rms multipath error *sE* versus target elevation *Et*, both normalized to radar beamwidth *qB*

The intermediate range is between the short-range region where the image appears in the sidelobes, and the long-range region, where the image appears within the half-power beamwidth. The error is difficult to calculate in this region because it falls in the nonlinear error-sensing portion of the antenna patterns, and the radar response is strongly dependent upon the specific feed design and error-processing technique. However, Fig. 4.14, provides a practical means for approximating multipath-error values in this region. The curves are calculated multipath errors based on an assumed gaussian-shaped sum pattern and derivative of the sum pattern as the monopulse difference pattern. Fig. 4.14 shows typical sidelobe multipath errors for higher-elevation targets and the linearly decreasing error versus target elevation, predicted by the above equation, for very low-elevation targets. The graph is normalized to radar beamwidth on both axes for convenient use with a wide variety of radars. The dashed portions of the curves are regions of uncertainty because of significant variations of reflection for a given sea state.

In the intermediate region, the error increases to a peak at target elevations of about 0.3 beamwidth. The peak value is dependent on several factors including surface roughness (which, in part, determines the value of ρ), servo bandwidth, and antenna characteristics in the region. The errors are severe, and with un-smoothed track (wide servo bandwidth), the radar can break lock and lose track of the target.

When the surface is rough, corresponding to a reflection coefficient of about 0.3, the characteristic of the error versus elevation changes is observed in Fig. 4.12. The rough surface causes significant diffuse scattering rather than a mirror reflection. This changes the shape of the error curve and results in some residual elevation-angle error when the target elevation goes to zero. It also causes some significant azimuth error

Crosstalk Caused by Cross-Polarized Energy. Target echo energy cross-polarized to the radar antenna causes crosstalk (cross coupling) in radars; *i.e.*, the azimuth error causes output from the elevation-error detector, and the elevation error causes output from the azimuth-error detector. Generally, this effect is negligible because the cross-polarized energy is usually a small fraction of the received polarization from typical targets, and it is normally reduced by about 20 dB by the antenna design. However, in special cases, the resultant crosstalk can be very high and may cause a large tracking error and possible loss of track. For example, polarization of a linearly polarized beacon on a target could rotate with target aspect change and, in the worst case, approach a cross-polarized condition.

Troposphere Propagation. The troposphere is typically a nonhomogeneous medium for propagation and will cause random beam bending. Figure 4.15 illustrates the approximate relation of rms angle error to various atmospheric conditions. The worst case is heavy cumulus clouds, which cause columns of air, shaded from the sun by the clouds, that are cooler than the surrounding air and consequently of a different dielectric constant. The result is typically a random beam bending as the radiated energy passes through these columns. Figure 4.15 applies only for the portion of the beam that is within the troposphere. Once the beam goes above the troposphere (typically about 6 to 9 km), there is no further beam bending.

The troposphere also affects target range measurement, but the errors are small, in the order of 0.3 to 0.6 m maximum. However, even small errors of this magnitude will cause significant errors in multilateration systems that determine target angle by calculations using precise range measurements from separate locations.

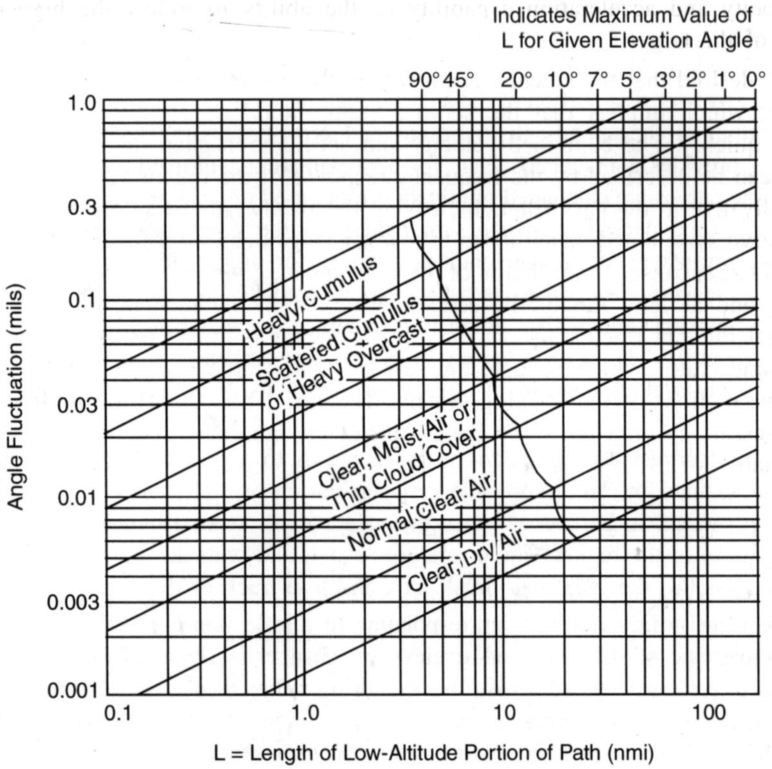

Fig. 4.15. Angle fluctuation versus path length for different tropospheres (from Final Report: Instrumentation Radar AN/FPS-16 (XN-2) by RCA under contract Bu Aer NOas 55-869c).

4.9. SERVO SYSTEMS FOR TRACKING RADAR

The servo system of tracking radar is the subsystem of the radar that receives as its input the tracking-error voltage and performs the task of moving the antenna beam in a direction that will reduce to zero the alignment error between the antenna axis and the target. For two-axis tracking with a mechanical-type antenna pedestal, there are typically separate axes of rotation for azimuth and elevation and separate servo systems to move the antenna about each axis. A conventional servo system is composed of amplifiers, filters, and a motor that moves the antenna in a direction to maintain the antenna axis on the target. Range tracking is accomplished by a similar system to maintain range gates centered on the received echo pulses. This may be accomplished by analog techniques or by digital-counter registers that retain numbers corresponding to target range to provide a closed range-tracking loop digitally.

Servo systems may use hydraulic-drive motors, conventional electric motors geared down to drive the antenna, or direct-drive electric motors where the antenna mechanical axis shaft is part of the armature, and the motor field is built into the support case. The direct drive is heavier for a given horsepower but eliminates gear backlash. Backlash may also be reduced with conventional motors by duplicate parallel drives with a small residual opposing torque when near zero angle rate. Amplifier gain and filter characteristics as well as motor torque and inertia determine the velocity and acceleration capability or the ability to follow the higher-order motion of the target.

It is desired that the antenna beam follow the center of the target as closely as possible, which implies that the servo system should be capable of moving the antenna quickly. The combined velocity and acceleration characteristics of a servo system can be described by the frequency response of the tracking loop, which acts essentially like a low-pass filter. Increasing the bandwidth increases the quickness of the servo system and its ability to follow a strong, steady signal closely. However, a typical target cause's scintillation of the echo signal, giving erroneous error-detector outputs, and at long range, the echo is weak, allowing receiver noise to cause additional random fluctuations on the error detector output. Consequently, a wide servo bandwidth, which reduces lag errors, allows the noise to cause greater erroneous motions of the tracking system. Therefore, for best overall performance, it is necessary to limit the servo bandwidth to the minimum necessary to maintain a reasonably small tracking-lag error. There is an optimum bandwidth that may be chosen to minimize the amplitude of the total erroneous outputs including both tracking lag and random noise, depending upon the target, its trajectory, and other radar parameters. The optimum bandwidth for angle tracking is range-dependent. A target with typical velocity at long range has low angle rates and a low SNR, and a narrower servo passband will follow the target with reasonably small tracking lag while minimizing the response to receiver thermal noise. At close range, the signal is strong, overriding receiver noise, but target angle scintillation errors proportional to the angular span of the target are large. A wider servo bandwidth is needed at close range to keep tracking lag within reasonable values, but it must not be wider than necessary or the target angle scintillation errors, which increase inversely proportional to target range, may become excessive.

The low-pass closed-loop characteristic of a servo system is unity at zero frequency, typically remaining near this value up to a frequency near the low-pass cutoff, where it may peak up to higher gain, as shown in Fig. 4.16(*a*). The peaking is an indication of system instability but is allowed to be as high as tolerable, typically to about 3 dB above unity gain to obtain maximum bandwidth for a given servomotor drive system. System A in Fig. 4.16(*a*) is a case of excessive peaking of about 8 dB. The effect of the peaking is observed by applying a step error input to the servo system. The peaking of the low-pass characteristic results in an overshoot when the antenna axis moves to align with the target. High peaking causes a large overshoot and a return to the target with additional overshoot. In the extreme (as in system A shown in Fig. 4.16(*b*), the antenna zeros in on the target with a damped oscillation. An optimum system compromise between speed of response and overshoot, as in system B, allows the antenna to make a small overshoot with reasonably rapid exponential movement back to the target. This corresponds to about 1.4 dB peaking of the closed-loop low-pass characteristic.

The resonant frequency of the antenna and servo system structure (including the structure foundation, which is a critical item) must be kept well above the bandwidth of the servo system; otherwise the system can oscillate at the resonant frequency. A factor of at least 10 is desirable for the ratio of system resonance frequency to servo bandwidth. High resonant frequency is difficult to obtain with a large antenna, such as the AN/FPQ-6 radar with a 29-ft dish, because of the large mass of the system.

The ratio was pushed to a very minimum of about 3 to obtain servo system bandwidth of the specified 3.5 Hz. Smaller radar with a 12-ft dish, for example, can provide a servo system bandwidth up to 7 or 8 Hz with conventional design.

However, because of this capability, the system can track multiple targets by rapidly switching from one to another rather than continuously tracking a single target.

The tracker simply places its beam at the location where the target is expected, corrects for the pointing error by converting error voltages (with known angle-error sensitivity) to units of angle, and moves to the next target. The system determines where the target was and, from calculations of target velocity and acceleration, predicts where it should be the next time the beam looks at the target. The lag error, in this case, is dependent on many factors, including the accuracy of the value of angle sensitivity used to convert error voltages to angular error, the size of the previous tracking error, and the time interval between looks.

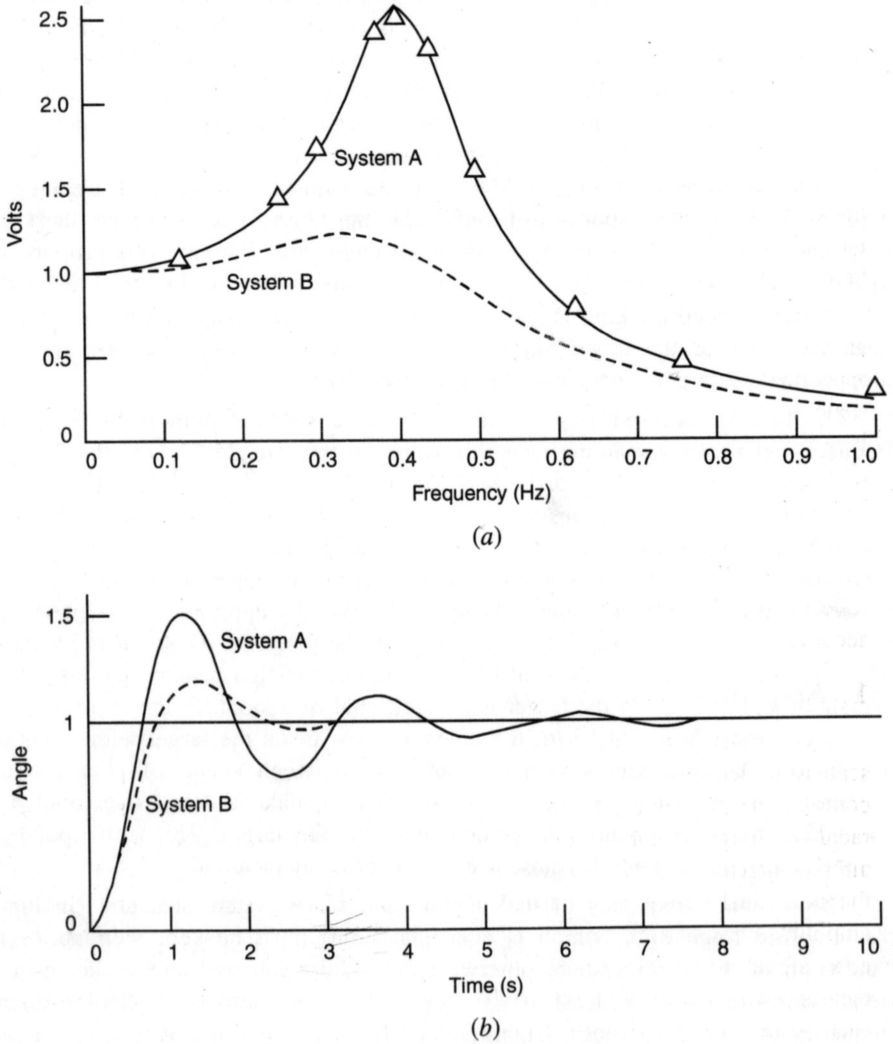

Fig. 4.16(*a*) Closed-loop frequency-response characteristics of two servo systems and (*b*) their corresponding time response to a step input.

4.10. LIMITATIONS TO TRACKING ACCURACY

The accuracy of tracking radar will be effected by such factors as the mechanical properties of the radar antenna and pedestal, the method by which the angular position of the antenna is measured, the quality of the servo systems, the stability of the electronic circuits, the noise level of receiver, the antenna beam width, atmospheric fluctuation, and the reflectors characteristics of the target. All these factors can degrade the tracking accuracy by causing the antenna beam to fluctuate in a random manner about the true target path. These noises like fluctuations are also known as *tracking noise or jitter.*

The above mentioned factors which affect the accuracy of tracking radar are summarized as:

1. Amplitude Fluctuation
2. Angle fluctuation or Target Glint
3. Receiver Noise
4. Servo noise
5. Atmospheric effects
6. Mechanical properties of antenna.

In many cases the two factors which ultimately limit the angular accuracy of practical tracking radars are the mechanical errors and the target reflectivity characteristics. The mechanical errors experienced in a precise monopulse tracking radar will not be discussed here.

A single radar target such as a smooth sphere will not cause degradation of the angular tracking accuracy. The radars cross-section of a sphere is impendent of the aspect at which it is viewed; consequently, its echo will not fluctuate with time. The same is true, in general, of a radar beacon if its antenna pattern is omni-directional. However, most radar targets are of a more complex nature then the sphere. The amplitude of the echo signal from a complex target may vary over wide limits as the aspect changes with respect to the radar. In addition, the effective center of radar reflection may also change.

Both of these effects, amplitude fluctuation and wandering of the radar center of reflection, as well as the limitation imposed by receiver noise can limit the tracking accuracy.

1. Amplitude Fluctuation

A complex target such as an aircraft or ship may be considered as a number of scattering elements. The echo signal can be represented as the vector addition of the contributions from the individual scatterer. If the target aspect changes with respect to radar—as might occur because of motion of the target or turbulence in the case of aircraft targets, the relative phase and amplitude relationship of the contributions from the individual scatter also change. Consequently, the vector sum, and therefore the amplitude change with changing target aspect. The amplitude fluctuation in the target echo signal which are also known as *target fading*, do not affect the angle error measurements accuracy of monopulse systems that extract an angle error voltage with each pulse.

Amplitude fluctuations of the echo signal are important in the design of the lobe-switching radar and the conical-scan radar but are of little consequence to the monopulse tracker radar. Both the lobe-switching tracker and the conical scan tracker require a finite time to obtain a measurement of the angle error. In the conical scan tracker, this time correspond to at least one revolution of the antenna beam and with lobe-switching, the minimum time is that necessary to obtain echoes at the four successive angular positions. In either case a minimum of four pulse repetition periods are required to make a measurement. If the target cross-section were to vary this observation time, the change might be errorously interpreted as an angular error signal.

The monopulse radar, on the other hand, determines the angular error on the basis of a single pulse. Its accuracy will therefore not be affected by changes in amplitude with time.

The echo signal from complex target is best described in statistical terms. Some of more useful statistical descriptions, the auto correlation function, the cumulative probability distribution, and the power spectral density. The power spectral density function is useful for describing the effect of amplitude fluctuation in the performance of a conical scan or lobe switching tracker.

To reduce the effect of amplitude noise on tracking, the conical scan frequency should be chosen to correspond to a low value of amplitude noise. If considerable amplitude fluctuation noise were to appear at the conical-scan or lobe-switching frequencies, it could not be readily eliminated with filters or AGC (automatic gain control).

A typical scan frequency might be of the order of 30 Hz. Higher frequencies might also be used since target amplitude noise generally decreases with increasing frequency. However, this may not always be true.

Propeller-driven aircraft produce modulation components at the blade frequency and harmonics thereof and cause a substantial increase in the spectral energy density at certain frequencies.

The scan frequency cannot be made higher than one quarter the pulse repetition frequency if a minimum of one hit per quadrant is to be obtained. It has been found experimentally that the tracking accuracy of radar operating with pulse repetitions frequencies from 1 kHz to 4 kHz and a lobing or scan rate one quarter of the PRF are not limited by echo amplitude fluctuation.

2. Angle Fluctuations or Target Glint

Changes in the target aspect with respect to the radar can cause the apparent center of radar reflection to wander from one point to another. The apparent center of radar reflection is the direction of the antenna when the error signal is zero. In general, the apparent center of reflection might not correspond to the target center. In fact, it need not to be confined to the physical extent of the target and may be off the target a significant fraction of time. The random wandering of the apparent radar reflecting center gives rise to noisy or jittered angle tracking. This form of tracking noise is called angle noise, angle scintillations, angle fluctuation or target glint.

The angular fluctuation produced by small target at long range may be of little consequence in most instances. However, at short range or with relatively large targets, angular fluctuations may be the chief limiting tracking accuracy. Angle fluctuations affect all tracking radars whether conical-scan, sequential lobing or monopulse tracking radar.

Glint is observed with complex target having many scattering center within the resolution cell of radar. These targets caused glint and degrade tracking. Glint from a complex target is sometimes thought of as a distortion of the echo wavefront.

To understand this affect, let us consider a simplified model of a complex radar target consisting of the independent isotropic scatterer's separated by an angular distance, θ_D, as measured from the radar. It is also a close approximation to the low angle tracking problem in which the radar sees the target plus its image reflected from the surface. The qualitative effects of target glint may be assessed from this model. The relative amplitude between the cross-section of the two scatterer's is assumed to be, *a*, and the relative phase difference is α. The differences in phase might be due to difference in range or to reflecting properties. The cross-section ratio, *a*, is defined as a number less than unity. The angular error Δθ as measured from the larger of the target is;

$$\frac{\Delta\theta}{\theta_D} = \frac{(a^2 + a\cos\alpha)}{(1 + a^2 + 2a\cos\alpha)} \quad ...(4.3)$$

The Fig. 4.17 shows the plot of Eq. (4.1). The position of the larger of the scatterers correspond to $\Delta\theta/\theta_D = 0$, while the smaller scatterer position is at $\Delta\theta/\theta_D = +1$. The positive values of Δθ relate to an apparent radar center which lies between the two scatterer's; negative values lie outside the target. When the echo signals from both scatterer's are in phase ($\alpha = 0$), the error reduces to $a/(a + 1)$, which corresponds to the so called " center of gravity" of the scatterers. When the signals are of equal amplitude ($a = 1$) and the phase difference, $\alpha = \pi$ radian, Eq. (4.1) indicates the antenna points to $\Delta\theta/\theta_D = -\infty$; that is, the antenna is driven well outside the bounds of the target.

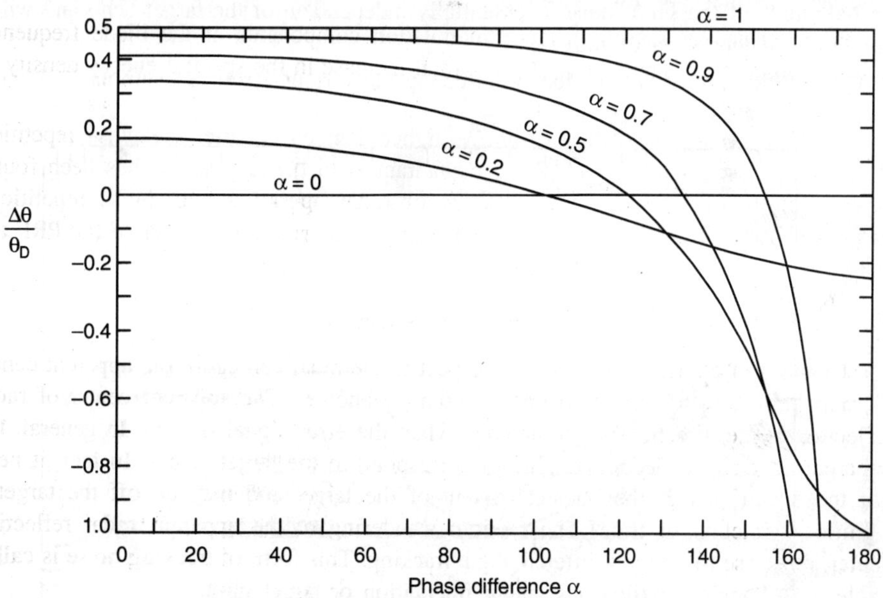

Fig. 4.17. Plot of Eq. (4.1).

Angle fluctuations are due to random changes in the relative distance from radar to the scatterer's that is varying of α. These changes may result from turbulences in the aircraft flight path or from the changing aspect caused by target motion. In essence, angle fluctuations are a distortion of the phase front of the echo signal reflected from a distortion of the phase front of the echo signal reflected from a complex target and may be visualized as the apparent tilt of this phase front as it arrives at the tracking system.

As per the Eq. (4.1), which indicates that the tracking error $\Delta\theta$ due to glint for the two scattering target is directly proportional to angular extent of the target θ_D. This is probably a reasonable approximation to the behaviour of real targets, provided the angular extent of the target is not too large compared with the antenna beamwidth. Since θ_D varies inversely with the distance for a fixed target size, the tracking error due to glint also varies inversely with distance.

Angle fluctuations in a tracking radar are reduced by increasing the time constant of AGC system (by reducing bandwidth), however his reduction in angle fluctuation is accompanied by new components of noise caused by amplitude fluctuations associated with echo signal; *i.e.*, narrowing the AGC bandwidth generates the additional noise in the vicinity of 0 frequency and poorer tracking results.

3. Receiver Noise

Receiver noise power is another limitation on tracking accuracy. The accuracy of the angle measurement is inversely proportional to the square root of SNR. Since the SNR is proportional to $1/R^4$ (from radar equation), the angular error due to receiver noise is proportional to the square of the target distance.

4. Servo Noise

The servo noise is due to hunting action of the tracing servo mechanism which results from backlash and compliance in the gears, shafts and structures of the mount. The magnitude of the servo noise is essentially independent of the target echo and will therefore be independent of range.

The contribution of various factors affecting the various errors are summarized and shown in Fig. 4.18.

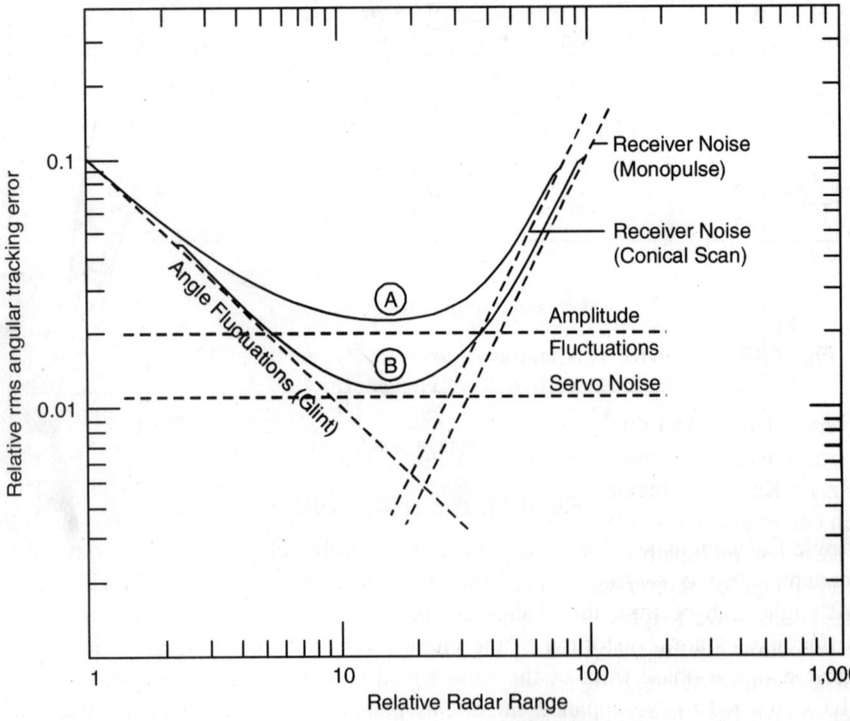

Fig. 4.18. Angular Tracking errors.

- angle fluctuations noise varies inversely with the range.
- receiver noise varies as the square of the range.
- amplitude fluctuations and servo noise are independent of noise.

4.11. ERROR REDUCTION TECHNIQUES

4.11.1. Multipath-Error Reduction

Very-low-altitude targets cause severe elevation-angle tracking errors, which may result in useless elevation tracking data and possible loss of tracking of the target. A variety of techniques have been developed to reduce these errors or their effects on radar tracking. One simple approach to avoid loss of tracking in elevation is to open the elevation-tracking servo loop and place the antenna beam at about a half beamwidth above the horizon. Azimuth closed-loop tracking may continue. Although the elevation-angle-error-detector output has large indicated angle errors, it is monitored to observe whether or not the target is maneuvering upward through the beam. A target rising through the beam will cause a positive angle-tracking error indication and the closed-loop elevation tracking resumes.

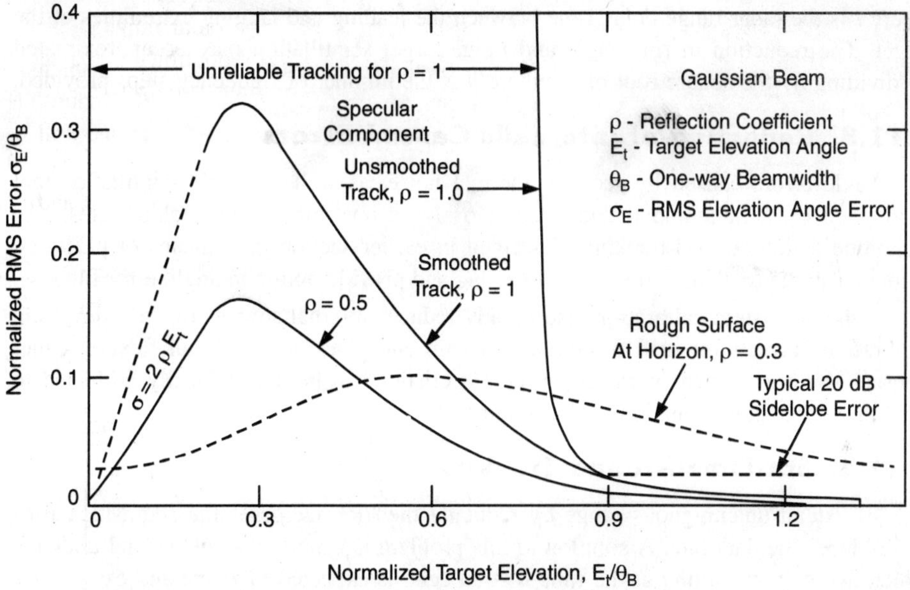

Fig. 4.19. Calculated rms multipath error σ_E versus target elevation E_t, both normalized to radar beamwidth θ_B.

A very effective and direct approach to multipath-error reduction is to use a very narrow beam, usually accomplished by operating at short wavelengths such as an 8-mm (35 GHz or Ka band) region with the usual microwave-tracking aperture size. This approach can reduce errors by two effects. First, as observed in Fig. 4.19, the magnitude of the elevation multipath error reduces in direct proportion to the beamwidth. The second advantage of shorter wavelengths is that even a relatively smooth sea, such as sea state 1, has wave heights of many wavelengths and appears rough, resulting in a smaller reflection coefficient. This is observed in Fig. 4.19 to give small multipath errors. The 8-mm-wavelength monopulse capability may be effectively combined with a lower microwave band to take advantage of the complementary features of both bands.

4.11.2. Target Angle and Range Scintillation (Glint) Reduction

Target-caused errors in angle and range tracking may be reduced by filtering, such as reducing tracking servo bandwidth. However, sufficient servo bandwidth must be retained to follow target trajectories. Unfortunately, target angle and range scintillation power density is normally concentrated below about 1 to 2 Hz when operating at microwave bands and falls within normally required bandwidths.

Target scintillation total noise power is relatively independent of frequency, but the spectral energy tends to spread upward in frequency as wavelength is reduced, resulting in lower noise power density in the servo passband. Therefore, operating at a shorter wavelength will result in lower target noise effects on closed-loop tracking.

Diversity techniques, which can provide statistically independent samples of target scintillation, offer a means for reducing target scintillation effects. The most practical technique is frequency diversity using pulse-to-pulse radar frequency change, which will alter the phase relations between the echoes from dominant reflecting surfaces of the target. The frequency change must be sufficient to cause enough change in relative phases of the reflectors to result in statistically independent samples of target scintillation at each new frequency. An approximate rule is a minimum frequency change of $1/t$, where t is the radar range delay time between the leading and lagging extremities of the target. The reduction in rms angle and range target scintillation may be approximated by dividing by the square root of n, where n is the number of frequency steps provided.

4.11.3. Reduction of Internally Caused Errors

Angle errors caused by receiver thermal noise, as well as target scintillation, are minimized by maintaining the target as closely as possible to the tracking axes. The technique called on-axis tracking, described in earlier section, is a means of placing a computer in the tracking loop to minimize lag and provide optimum angle-error filtering.

Accurate system calibration also greatly reduces internal sources of error. Frequent calibration corrects for drift in component gain and phase and pedestal flexure. Other internal sources of error with known characteristics can be automatically corrected to minimize their contamination of the output data.

4.11.4. nth-Time-Around Tracking

To extend unambiguous range by reducing the PRF increases the acquisition time and reduces the data rate. A solution to this problem is called *n*th-time-around tracking, which avoids transmitting at the time that an echo is expected to arrive and can resolve the range ambiguity. This allows the radar to operate at high PRF and track unambiguously to long ranges where several pulses may be propagating in space to and from the target. The technique is useful only when a target is being tracked. During acquisition, the radar must look at the region between transmitter pulses, and upon initial acquisition, it closes the range- and angle-tracking loops without resolving the range ambiguity. The next step is to find which range interval, or between which pair of transmit pulses, the target is located. The zone n is determined by coding a transmit pulse and counting how many pulses return before the coded pulse returns.

Instrumentation radars provide nth-time-around tracking capability because beacons are used on rockets and space vehicles to provide sufficient signal level at very long ranges.

To prevent the target echo from being blanked by a transmit pulse, it is necessary to sense when the target is approaching an interference region and shift the region. This is accomplished by changing the PRF or alternately delaying groups of pulses equal to the number of pulses in propagation. This can be performed automatically to provide an optimum PRF shift or to alternately delay pulse groups of the correct number of pulses.

4.12. MIRROR-SCANNED ANTENNA (INVERSE CASSEGRAIN)

An antenna technique that uses a movable RF mirror for scanning the beam, called a mirror-scanned antenna or inverse Cassegrain, provides useful applications to monopulse radar. The technique uses a radome-supported wire-grid paraboloid that reflects parallel-polarized feed energy. The beam, polarized parallel to the grid, is collimated by the paraboloid and is reflected by a flat moveable polarization rotating mirror. The basic polarization rotating mirror is a flat metal surface with a grid of wires located a quarter wavelengths above the metal surface and oriented at 45° relative to the RF energy reflected from the paraboloid. The RF energy may be visualized as being composed of a component parallel to and reflecting from the grid and a component perpendicular to and passing through the grid to reflect from the metal mirror surface below. By travelling the quarter wavespace twice, this component is shifted by 180° in phase. When added to the reflection from the grid, it results in a 90° change in polarization. The total reflected energy from the mirror rotated by 90° will efficiently pass through the wire-grid paraboloid.

The advantages are as follows:

(1) The mirror and its drive mechanism are the only moving parts for beam movement. The feed and radome-supported paraboloid remains fixed.

(2) The beam movement is by specula reflection, twice the angle of the mirror tilt. This provides a compact structure for a given angle coverage requirement.

(3) The normally lightweight mirror and the 2:1 beam displacement versus mirror tilt allow reduced size and very rapid beam scan with low servo drive power.

The compactness and lightness are particularly attractive for airborne applications such as the Thompson-CSF Agave radar in the Super Entendards, which determines target range and designation data for the Exocet missile. It is compact monopulse roll- and pitch-stabilized radar with 140° azimuth and 60° elevation scan. The Israeli Elta subsidiary of Israeli Aircraft Industries also developed an airborne tracking radar using this antenna technology for air-to-air combat and ground weapon delivery.

A ground- or shipboard-based experimental mirror antenna system concept was developed with dual-band monopulse capability (3.0 GHz and 9.3 GHz bands). The objective included high-speed beam movement for high-data-rate 3D surveillance and multitarget precision tracking. Dual-band polarization-twist mirror design was accomplished with a two-layer mirror grid configuration.

4.13. ON-AXIS TRACKING

The best radar tracking performance is usually accomplished when the target is essentially on the radar antenna axis. Therefore, for maximum precision tracking, it is desirable to minimize lag and other error sources affecting the beam pointing. A technique called on-axis tracking was developed to minimize radar axis deviation from the target by prediction and optimum filtering within the tracking loop. The technique is particularly effective when the target trajectory is known approximately, such as when tracking satellites in orbit or a ballistic target. A computer in the tracking loop can cause the radar to follow an estimated set of orbital parameters, for example. It also performs optimum filtering of radar angle-error-detector output to generate an error trend from which it can update the assumed set of orbital parameters to correct the radar beam movement to update the original set of orbital parameters, and by this means, the radar antenna axis can be held on target with minimum error.

Improved tracking can also be provided on other targets where the approximate trajectory can be anticipated. However, performance of on-axis tracking is limited when tracking targets with unanticipated maneuvers.

4.14. COMPARISONS OF TRACKING RADAR

As we have discussed many techniques of tracking radars which are as follows;

1. Conical-scan radar
2. Sequential-lobing radar
3. Amplitude-comparison Monopulse tracking radar
4. Phase-comparison Monopulse tracking radar.

The conical scan and amplitude-comparison monopulse tracking techniques have more number of applications. The phase-comparison monopulse tracking technique is not in more use because of four separate antennas, which point their beams in the same direction and the sidelobes are high.

Conical scan has very low less, and the antenna and feed systems are less complex.

The signal to noise ratio in monopulse radar is more than that of conical scan radar.

The difference in SNR between the monopulse and the conical scan radars is about 2 dB to 4 dB.

The beamwidth in the conical scan radar antenna is slightly greater than that of monopulse radar because it uses offset feed from the focus.

The monopulse radar views the target at peak of sum pattern whereas conical scan radar views the target at an angle off the peak of the antenna beam.

The tracking accuracy is superior in monopulse radar and that is why it is preferred for precision tracking. In the presence of glint, the tracking radar monopulse and conical scan radars have poor tracking accuracy.

Monopulse radar is more complex than conical scan radar. Monopulse radar required three separate receivers whereas conical radar uses only one receiver.

In monopulse radar, angle measurement is made first and then the number of pulses are integrated to obtain SNR and to smooth the error. In conical scan radar, the number of pulses is integrated first and then the angle is measured.

Monopulse radar is not affected by amplitude fluctuations and hence it is less susceptible to enemy Electronic Counter Measure (ECM) than the conical scan radar.

We can summarize the comparison of monopulse and conical scan tracking radar in the following table:

S. no.	Monopulse Radar	Conical scan Radar
1.	Range accuracy is better.	Range accuracy is inferior to the monopulse radar.
2.	Angle accuracy is better.	Angle accuracy is inferior to the monopulse radar.
3.	SNR is more.	SNR is less.
4.	Angle accuracy is not affected by fluctuation in the echo amplitude.	Angle accuracy is not affected by fluctuation in the echo amplitude.
5.	Glint, degraded the system performance.	Glint also degraded the system performance.
6.	Monopulse radar is more complex.	Conical scan is less complex.
7.	It is simultaneous tracking system.	It is sequential tracking system.

4.15. TRACKING WITH SURVEILLANCE RADAR

The track of a target can be determined with surveillance radar from the coordinates of the target measured from scan-to-scan. The quality of such a track depends on the time between observations, the location accuracy of each observation and the number of extraneous targets present in the vicinity of the tracked target. A surveillance radar which develops tracks on targets it has detected is called a "*track while scan*" (TWS). Radar tracks can be obtained by having an operator mark the location of a target on the face of the PPI with a grease pencil on each scan. A single operator however cannot handle more than about 6 target tracks when the radar has a twelve second scan rate also an operator's effectiveness in detecting new targets decreases rapidly after 1/2 hour of operation. These problems are avoided by automating the target detection and tracking process. This is called "automatic detection and tracking (ADT)".

The ADT performs the following functions:

1. Target detection
2. Track initiation
3. Track association
4. Track update
5. Track smoothing
6. Track termination.

The automatic detection quantizes the range into intervals equal to the range resolution. At each range interval the detector integrates n pulses (n is the expected number of hits expected from the target due to the antenna scan rate). The integrated pulses are compared with the threshold to determine the presence or absence of a target. An example is the "moving window detector" which examines continuously. The last n samples within each quantized range and announces the presence of a target. If m out of n of these samples crosses a preset threshold by locating the centre of the n pulses, an estimate of the target's angular direction can be obtained. This is called "beam splitting" if there is only one target present within the radar's coverage, then the detection on two scans is all that is required to establish a target track and estimate its' velocity.

However there are usually other targets plus clutter echoes and hence three or more detections are necessary to establish a track reliably without the generation of false tracks. A computer can recognize and reject false tracks, however too many false tracks can overload a computer. Hence the radar using ADT should exclude unwanted signals from clutter and interference. A good ADT system therefore requires a good MTI and a good CFAR receiver. A clutter map generated by the radar is sometimes used to reduce the load on the tracking computer by blanking the clutter areas and removing detections associated with large point clutter sources not rejected by the MTI. Slowly moving echoes that are not of interest can also be removed by the clutter map. The availability of distinctive target characteristics such as altitude might prove of help when performing track association. When a new detection is received, an attempt is made to associate it with existing tracks this is aided by establishing for each track. A small search region (gate) within which a new detection is predicted, based on the estimate of the target speed and direction.

It is desired to make the gate as small as possible to avoid having more than one echo fall within it when traffic density is high, or when two tracks are close. However, a large gate is required if the tracker is to follow target maneuvers. Hence more than one gate size may be used. The size of the gate depends on the accuracy of the track. When a target does not appear in the small gate, a larger gate could be used. The size of the second gate would depend on the maximum acceleration expected of the target on the basis of past detections, the track while scan radar must make smoothed estimates of the target position and velocity as well as the predicted position and velocity.

One method of computing this information is α-β tracker. Here the present smoothed target position $\bar{x}$ and $\bar{v}$ are computed using

$$\bar{x}_n = x_{pn} + \alpha(x_n - x_{pn})$$

$$\bar{v}_n = \overline{v_{n-1}} + \frac{\beta}{T_s}(x_n - x_{pn})$$

where, X_{pn} = the predicted position of the target at the nth scan

X_n = the measured position at the nth scan

α = the position smoothing parameter

β = the velocity smoothing parameter

T_s = the time between observations

The predicted position at the $n + 1$ scan is

$$x_{pn+1} = \bar{x}_n + \bar{v}_n T_s$$

When acceleration is important, a third equation can be added to describe a α–β–γ tracker,

where, γ is the acceleration smoothing parameter

- if $\alpha = \beta = 0$ then the tracker uses no current data, only smoothed data.
- if $\alpha = \beta = 1$ then no smoothing is included at all.

The α–β filter is designed to minimize the mean square error in the smoothed (filtered) position and velocity. Assuming small velocity changes between observations to minimize the output noise variance at steady state, and the transient response to a maneuvering target modeled by a ramp function. The α–β coefficients are related by

$$\beta = \frac{\alpha^2}{2-\alpha}$$

The choice of α between 0 and 1 depends on the system application and in particular on the tracking bandwidth a compromise must be made between good smoothing (narrow bandwidth) and rapid response to maneuver (wide bandwidth) a criterion for selecting the α, β coefficients is based on the best linear track fitted to the radar data in the least squares sense. This gives values

$$\alpha = \frac{2(2n-1)}{n(n+1)} \qquad n > 2$$

$$\beta = \frac{6}{n(n+1)} \qquad n > 2$$

where, n = the number of scans or target observations.

A standard α – β tracker does not handle the maneuvering target.

An adaptive α – β tracker is one which varies the two smoothing parameters to achieve a variable bandwidth so as to be able to follow maneuvers.

The value of α can be set by observing the measurement error $x_n - x_{pn}$.

At the start of the tracking, the bandwidth is set to be wide and is narrowed down if the target moves in a straight line. As the target maneuvers, the bandwidth is widened to keep the tracking error small.

Kalman filter is similar to an adaptive α – β tracker except that it inherently provides for dynamical targets. Here a model for the measurement error has to be assumed, as well a model of the target trajectory and the disturbance of the trajectory. Disturbances might be due to neglect of higher order derivatives in the model of the dynamics, random motions due to atmospheric turbulence, and target maneuvers. Kalman filters can use a wide variety of models for measurement noise and trajectory disturbance, however these are usually assumed to be white noise with zero mean a maneuvering target does not always fit this model since it produces correlated observations. Proper inclusion of realistic dynamical models increases the complexity of the calculations.

The Kalman filter is sophisticated but costly to implement, its chief advantage over the classical α – β tracker is its inherent ability to account for maneuver statistics.

If Kalman filters were restricted to modelling the target trajectory as a straight line. and if the measurement noise and trajectory disturbance noise were modeled as white Gaussian noise with zero mean, then the Kalman filter equations reduce to α – β tracker equations, with the α and β computed sequentially by the Kalman filter procedure. The classical α – β is easy to implement to handle maneuvering targets, some means maybe included to detect maneuvers and to change the values of α and β.

The data rate might also be increased during target maneuvers in some radar systems. As means for choosing α and β become more sophisticated, the optimal α – β tracker becomes equivalent to a Kalman filter even for a target trajectory with error here. The computation of α and β require a knowledge of the statistics of the measurement errors and the prediction errors, and are determined in a recursive manner in that they depend on previous estimates of the mean square error in the smoothed position and velocity.

Note: The concept of the α – β tracker or the Kalman filter apply also to a continuous single target tracker where the error signal is processed digitally. The equations describing the α – β tracker are equivalent to a Type II servo system of the track while scan radar, does not receive target information on a particular scan, the smoothing and predicting operations can be continued by accounting for the missing data. When data to update a track is missing for a sufficient number of consecutive scans, the track is terminated.

Example: If the criterion for establishing a track is 3 target reports, 5 consecutive misses is suitable for termination.

ADT effectively reduces the bandwidth in the radar output allowing the data to be transmitted to a remote site via narrow band phone lines. This allows the outputs from several radars to be communicated to a central control point economically. The adaptive thresholding of the automatic detector can cause worsening of the range resolution. It would seem that two targets might be resolved if their separation is about 0.8 pulse width.

However with automatic detection, the probability of resolving targets in range only exceeds 0.9 if they are separated by 2.5 pulse widths.

To achieve this resolution, a log video receiver should be used and the threshold should be proportioned to the smaller of the two means calculated from a number of reference cells on either side of the test cell. This resolution limit assumes that the shape of the return pulse is not known. (If it were, it would be possible to resolve targets within one pulse width).

When more than one radars are covering the same volume of space, it is sometimes desirable to combine their outputs to form a single track file (Automatic Detection and Integrated Tracking, ADIT). ADIT has the advantage of a greater data rate than single radar and reduces the likelihood of a loss of target detections caused by antenna lobing, fading, interference and clutter. The integrated processing allows a favourable weighting of better data and lesser weighting of the poorer data.

SUMMARY

- A tracking radar system measures the different co-ordinate of a target such as range, elevation angle, azimuth angle, and Doppler shift frequency to determine the target path and to predict its future position.
- One of the most basic tracking system designs is the servo tracking system.
- There are various methods for generating the error signal may be classified as given below:
  1. **A sequential lobbing :** A single beam is switched between two sequential angular positions to obtain an angle measurement. This is called sequential lobbing, lobe switching or sequential switching.
  2. **A conical scan:** In the conical scanning, the squinted beam is scanned rapidly and continuously in a circular path around the axis. The angle between the axis of rotation and the axis of the antenna beam is called the **squint angle**. If a target is present within squint angle, then the echo signal from the target will be amplitude modulated at a frequency equal to the rotation frequency of the beam also called conical scan frequency.
  3. **A mono pulse.** The tracking techniques that derive angle error information on the basis of single pulse, is known as **mono pulse tracking or simultaneous lobbing**. There are several methods by which a mono pulse angle measurement can be made.

     (*a*) Amplitude Comparison Mono Pulse

     (*b*) Phase Comparison Mono Pulse Tracking

- **Acquisition radar.** When separate search radar employed for giving the information of the target to the tracking radar system, is called acquisition radar.
- **Types of Tracking Radar**
  1. Single target tracker (STT)
  2. Automatic detection and track (ADT)
  3. Phased array radar tracking (multiple target tracking)
  4. Track while scan. (TWS)
- **Tracking in Range :**
  1. Split-gate-tracker
  2. Range Glint
- **Acquisition.** A tracking radar must first find and acquire its target before it can operates as tracker.
- There are different types of scanning pattern employed with pencil beam antennas
  1. Helical scanning beam
  2. Palmer scan
  3. Spiral scan
  4. Raster or TV scan
  5. Nodding scan
- **Synthetic aperture radar (SAR),** Synthetic aperture radar (SAR) is a form of radar in which sophisticated post-processing of radar data is used to produce a very narrow effective beam.
- $$\Delta R_{cross} = R \cdot (k\lambda/D)$$
  $$\Delta R_{cross} = R \cdot \lambda/D$$
  If the length of the SAR antenna is (L) than beamwidth will be
  $$\theta = k.\lambda/2L$$
- **Ultra-wideband SAR.** Ultra-wideband radar emits very short pulses consisting of a very wide range of frequencies, from zero up to the radar's normal operating frequency. Such pulses allow high distance resolution but much of the information is concentrated in relatively low frequencies (with long wavelengths). Thus such systems require very large receiving apertures to obtain correspondingly high resolution along the track. This can be achieved with synthetic aperture techniques.
- **Interferometry.** Rather than discarding the phase information, information can be extracted from it. If two observations of the same terrain from very similar positions are available, a great deal of interesting information can be extracted. This technique is called interferometric SAR or ISAR.
- **Inverse Synthetic Aperture Radar (ISAR).** It is possible to achieve the same large synthetic aperture without moving the transmitter/receiver. If the target rotates by a small amount, it has the same effect as if the transmitter/receiver were to travel a distance equal to the arc length at the range.
- **Application of SAR.** It has seen wide applications in remote sensing and mapping. The imaging of the earth surface by SAR to provide a map like display can be applied to millitary reconnaissance, measurement of sea state and ocean wave condition, geological and mineral.
- The accuracy of tracking radar will be effected by such factors as the mechanical

properties of the radar antenna and pedestal, the method by which the angular position of the antenna is measured, the quality of the servo systems, the stability of the electronic circuits, the noise level of receiver, the antenna beam width, atmospheric fluctuation, and the reflectors characteristics of the target.

- In many cases the two factors which ultimately limit the angular accuracy of practical tracking radars are the mechanical errors and the target reflectivity characteristics.
- The amplitude fluctuation in the target echo signal which are also known as *target fading*, do not affect the angle error measurements accuracy of monopulse systems that extract an angle error voltage with each pulse.
- Amplitude fluctuations of the echo signal are important in the design of the lobeswitching radar and the conical-scan radar but are of little consequence to the monopulse tracker radar.
- Changes in the target aspect with respect to the radar can cause the apparent center of radar reflection to wander from one point to another.
- Angle fluctuations affect all tracking radars whether conical-scan, sequential lobing or monopulse tracking radar.
- Glint is observed with complex target having many scattering center within the resolution cell of radar. These targets caused glint and degrade tracking.
- Receiver noise power is another limitation on tracking accuracy.
- The servo noise is due to hunting action of the tracing servo mechanism which results from backlash and compliance in the gears, shafts and structures of the mount.
- Conventional monopulse radar tracking equipment suffers severe degradation of tracking accuracy when tracking targets at low angles.
- Conventional monopulse radar systems are incapable of distinguishing between the direct reflected signal and its image signal, creating ambiguity in the display.
- Multipath angle errors result from reflections of the target echo from objects or surfaces causing echo pulses to arrive by other than the direct path to the radar beam in addition to the direct path.
- The conical scan and amplitude-comparison monopulse tracking techniques have more number of applications. The phase-comparison monopulse tracking technique is not in more use because of four separate antennas, which point their beams in the same direction and the sidelobes are high.
- The track of a target can be determined with surveillance radar from the coordinates of the target measured from scan-to-scan.

REVIEW QUESTIONS

1. What do you understand by term Tracking Radar ?
2. Explain the radar servo tracking system with the help of simple block diagram.
3. Explain various types of tracking radar ?
4. Describe the sequential lobbing method of tracking of an acquired target.
5. Describe the conical scanning method of tracking of an acquired target. How is this an improvement over lobe switching ?
6. Explain the working of mono pulse tracking radar for one angle coordinates.

7. Explain the working of mono pulse radar with the help of a block diagram. **[UPTU, 2003]**
8. Describe the various methods of mono pulse tracking technique.
9. Define tracking in range and explain split gate tracker method.
10. What do you understand by acquisition of target and explain the various method of antenna scanning ?
11. What are the various Antenna scanning ? Explain each in detail.
12. Explain the working principle of Synthetic Aperture Radar (SAR).
13. Write the application of SAR.
14. Describe the Inverse Synthetic Aperture Radar (ISAR).
15. What are the various techniques of tracking ?
16. Explain the basic difference search radar and tracking radar.
17. Compare different types of trackers system.
18. With the aid of a sketch, describe the equipment and technique used in monopulse method of target tracking ?
19. How is conical scanning an improvement over lobe switching ?
20. Explain the block diagram of amplitude comparison in monopulse in one angle coordinate.
21. Explain the block diagram of conical scan tracking radar.
22. Explain split gate tracker and range glint.
23. Briefly discuss the OTH radar and write its applications.
24. What do you understand by limitations to tracking accuracy?
25. Explain various types of limitations to tracking accuracy.
26. Explain Amplitude fluctuation and how it affects the performance of tracking radar.
27. Discuss target glint and how it affects the performance of tracking radar.
28. Explain low angle tracking in details.
29. Give the comparison of various trackers.
30. Discuss automatic tracking with surveillance radar.
31. Write shorts on :
    (*a*) Sequential lobbing
    (*b*) Conical scanning
    (*c*) Amplitude comparison mono pulse tracking
    (*d*) Phase comparison mono pulse tracking
    (*e*) Range Glint
    (*f*) Split-gate tracker
    (*g*) Helical scanning
    (*h*) Palmer scanning
    (*i*) Nodding scanning
    (*j*) Ultra-wideband SAR
    (*k*) Interferometery
    (*l*) OTH Radar
32. Explain the following in detail :
    (*a*) Sequential Lobbing
    (*b*) Monopulse Tracking Radar.

❑❑❑

CHAPTER

5

DETECTION OF SIGNALS IN NOISE

INSIDE THIS CHAPTER

5.1. INTRODUCTION

In this chapter different aspects of the Detection of radar signal in noise are discussed. The radar has to perform two basic functions such as detection of the signal in the noise and second is extraction of target information in the term of position, velocity and size of the target. A radar which has good detection capabilities may have the good extraction ability also and vice-versa.

This chapter will cover the different aspect of detection such as the matched filter receiver, the cross correlator detection, and detection criteria. This chapter is divided in to two sections, the first section deals with optimization of receiver to detect the weak signal in noise and in the second section the study of many detection criteria to decide whether the output of the radar receiver is due to noise alone or it is signal with the noise. The detection criteria are based upon the statistical testing such as Neyman-Pearson, Ideal Observer, and the Sequential tests.

5.2. MATCHED FILTER RECEIVER

The matched–filter receiver is an optimum method for detection of reflected signals in noise. A characteristic of the radar receiver which makes the frequency response function maximize to the output peak signal-to-mean-noise ratio is called a **matched filter**. It is equal to the $2E_0/N_0$, Where E_0 is the received signal energy and N_0 is noise power in the bandwidth.

When the input to the network is purely sinusoidal, than the frequency response function, denoted by $H(f)$, describe the relative amplitude and phase of the output of that network with respect to the input. The magnitude of the frequency response function is denoted by $| H(f) |$ and is relative to the band pass characteristics of the receiver. Wider the bandwidth, lower the output S/N ratio because of extraneous noise is introduced by the excess bandwidth. In case of narrow bandwidth again reduces the output S/N ratio because it limits the signal along with the noise. Therefore the bandwidth of the radar receiver is so chosen that the loss of signal should be minimum and output S/N ratio must be at optimum. There is one thumb rule in pulse radar that the bandwidth of receiver should be approximately equal to the reciprocal of the pulse width (T), it can be written as

$$BW \approx 1/T \qquad ...(5.1)$$

The above perception is true for the pulse radar using superhetrodyne receiver but it may not be true for other shapes of waveform. Generally, the optimum receiver characteristics depend upon the frequency response function and the shape of the received waveform.

The receiver frequency response function applied from the antenna to the output of IF Amplifier. The bandwidth of the RF and mixer stage is larger than the bandwidth of the IF Amplifier stage. Therefore, the narrow banding is most convenient at the input of the IF amplifier. So it is necessary to consider the frequency response function that maximize the S/N ratio at the output of IF amplifier. The IF amplifier act as the filter with the gain and the characteristics of this filter is essential to design the optimum matched filter.

The frequency response function of the linear, time invariant filter which maximize the output peak signal-to-mean-noise (power) ratio for a fixed input S/N (energy) ratio is given below in Eqn. (5.2), if the received waveform is $s(t)$ with a given ratio of signal energy E_0 to noise energy N_0.

$$H(f) = G_a \times S^*(f) \times \exp(-j2\pi f t_1) \qquad ...(5.2)$$

$$S(f) = \int_{-\infty}^{\infty} S(t) \exp(-j2\pi f t)\, dt$$

where,
$S(f)$ = Voltage spectrum (fourier transform) of I/P signal
$S^*(f)$ = complex conjugate of $S(t)$
t_1 = The value of time at which signal is observed
G_a = Constant equal to maximum filter gain (unity)

It is assumed that the noise accompanies the signal are white noise (stationary and have uniform spectrum). The noise need not be Gaussian.

The frequency response given in the above Eqn. (5.2) is known as the North Filter, conjugate filter or usually the matched filter. Some time it may be called as the Fourier transform criterion.

The above frequency response function of the matched filter does not consider the phase shift $\exp(-j2\pi ft)$. This phase shift will change uniformly according to frequency. A constant time delay can achieve with the phase shift. The matched filter should not give the output until the signal is applied to it, and can be achieved by a time delay. The frequency spectrum of the receiver may be written as a phase spectrum $\exp[-j\,\Phi_s(f)]$ and an amplitude spectrum $[| S(f) |]$. Similarly it may be written in terms of its amplitude and phase spectra $| H(f) |$ and $\exp[-j\,\Phi_m(f)]$ respectively. Ignoring the G_a constant and Eqn. (5.2) can be re-written for the matched filter as

$$| H(f) | \times \exp[-j\Phi_m(f)] = [| S(f) |] \exp[j\{\Phi_s(f) - 2\pi f t_1\}] \qquad ...(5.3)$$

or

$$| H(f) | = [| S(f) |] \qquad ...(5.4)$$

Where $\Phi_m(f) = -\Phi_s(f) + 2\pi f t_1$ ...(5.5)

By analyzing the above Eqn. (5.3), (5.4) and (5.5), we can say that the amplitude spectrum of the matched filter is equal to the amplitude spectrum of the received signal but the phase spectrum of the matched filter is the negative of the phase spectrum of the signal and sum of phase shift, which is proportional to the frequency.

In term of impulse response $h(t)$, which is inverse of the Fourier transform of the frequency –response function, the matched filter may written as

$$h(t) = \int_{-\infty}^{\infty} H(t) \exp(j2\pi f t) df \qquad ...(5.6)$$

The impulse response is the output of the matched filter as a function of the time when input is an impulse *i.e.,* delta function. Substituting the Eqn. (5.2) in to Eqn. (5.6), we get

$$h(t) = G_a \int_{-\infty}^{\infty} S^*(f) \exp[-j2\pi f(t_1 - t)] df \qquad ...(5.7)$$

Since $S^*(f) = S(-f)$

We get, $$h(t) = G_a \int_{-\infty}^{\infty} S(f) \exp[j2\pi f(t_1 - t)] dt \qquad ...(5.8)$$

or $$h(t) = G_a \cdot S(t_1 - t) \qquad ...(5.9)$$

The impulse response of the matched filter is the image of the received waveform as shown in Fig. 5.1, a received waveform $S(t)$ and the impulse function $h(t)$ of its matched filter.

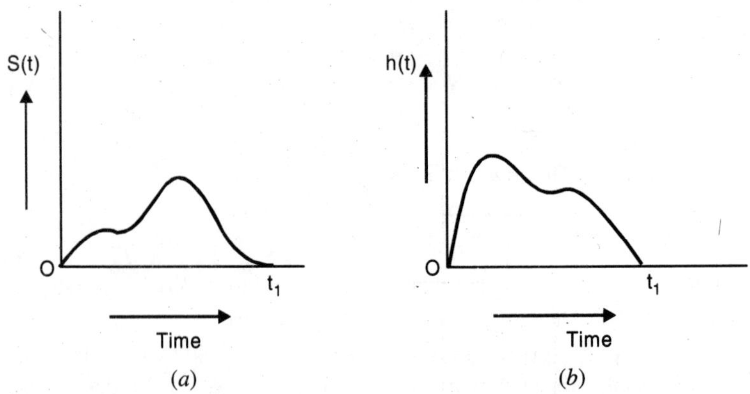

Fig. 5.1. **(*a*) and (*b*) Shows a Received Wave form (*a*) and Impulse Response (*b*).**

5.3. EFFECT OF S/N RATIO OVER THE DETECTION OF WEAK SIGNAL

The matched filter defined as that which maximize the output—Peak signal to Mean Noise Ratio. Obviously, that filter should be the best filter to use to maximize the weak signal detection provided the matched filter response is linear in nature. If the S/N ratio is increased by the non-linear devices in the receiver then the detection of weak signal might not increased evenly enhancing the S/N ratio. So the maximum signal to noise ratio will not work out, if the matched filter receiver device not linear.

5.4. CORRELATION DETECTION

This brings up the question of how the time of arrival of the reflected signal is determined. The basis of this determination involves the computation of the correlation between the two signals, the outgoing and incoming. Correlation is a measure of the similarity or relatedness between two waveforms. For two waveforms v_1 and v_2 the correlation is mathematically given by the following equation

$$R(t) = \int_{-T/2}^{T/2} v_1(t')\, v_2(t+t')\,dt' \qquad ...(5.10)$$

If $T \rightarrow \infty$

If they have the same fundamental period T_0, then T can be replaced by the same and the average cross correlation can be computed. It should be noted that the correlation depends on the time shift, t given to the waveforms. This shift results in a maximum correlation at some points and zero at others. Two waveforms are considered to be coherent if they are related while they are uncorrelated or incoherent if there is no match between them at any given time. Auto correlation is the measure of the coherence of a waveform with itself. It is noted that in the case of auto correlation it would be maximum when the time shift would be zero or a multiple of its time period. In radar signalling the received and sent signal are basically the same so it is the auto correlation that is computed. The two signals *i.e.* the one broadcast and the one received are matched continuously. The instant when they match or the auto correlation is the maximum is the point at which the signal is considered to have arrived. This computation is done with the help of a matched filter.

Correlation is the **optimal** technique for detecting a known waveform in random noise. That is, the peak is higher above the noise using correlation than can be produced by any other linear system. Using correlation to detect a known waveform is frequently called **matched filtering**.

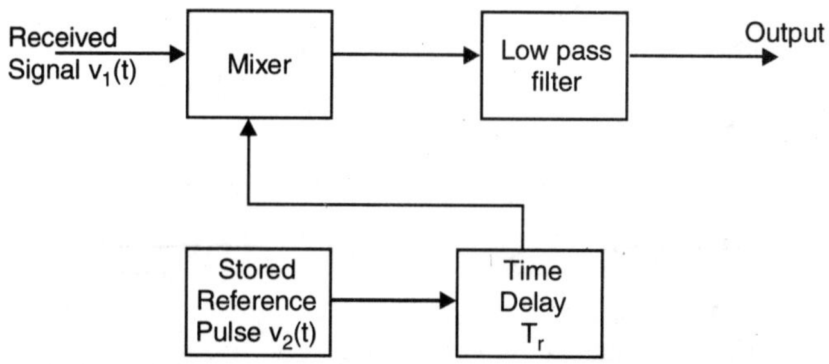

Fig. 5.2. Block Diagram of Cross Correlation Detector.

Cross correlation is measure of the coherence between two pulses.The principle of the cross correlation function may be used as a basis for radar receiver design and can be explained with the help of Fig. 5.2. Assumed that $v_1(t)$ to be the input signal to the receiver, it consists of a pulse of sine wave along with noise. The frequency and duration of of input pulse are known. The reference waveform $v_2(t)$, is stored in the receiver and it is similar to the expected signal as $v_1(t)$. Since the noise waveform that accompanies input pulse is random and unknown to the receiver, so it is better to compute cross correlation rather than autocorrelation function. The receiver computes the cross correlation function for various values of T_r (the time delay to determine the range).

If the magnitude of $R(t)$, correlation function exceeds a predetermined threshold level for various values of T_r than the target is said to be detected at a range $R = cT_r/2$.

5.5. THRESHOLD DETECTION CRITERIA

It is the main interest of radar engineer to know that the received signal is either from the noise or from the target. The detection of the weak signal in the presence of noise to decide whether the received signal is due to noise or noise along with signal. The human operator takes the decision on the basis of information available at the radar display. In case of automatic detection without the aid of operator, must be carefully specified and integrate the decision making circuits by the radar engineer.

Usually the radar detection is based upon the threshold detection, in that o/p of the receiver is compared with the predefined threshold level. If the level of signal crosses the threshold level, a signal is said to be target and below the threshold level is said to have noise. The threshold level is to define the received signal in this category either it is present or not.

Now, the threshold level of the receiver has the significant aspect of the detection. In the first case, if the radar designer keeps the level of threshold too low, then there are chances of getting false echoes over the screen, which makes difficult to detect the target from the noise for the radar observers. In radar engineering, this term is known as false alarm and in statistical detection theory it is some time known as type I error.

In second case, if the radar engineers choose the threshold level is large enough, than there are chances of missing detection of the actual targets. These are being treated as noise but actually they are desired targets. So, in this the radar observer miss the desire targets because of weak signal or the threshold level is so high to cross that. This is a missed detection for the radar engineer and type II error in statistical detection theory.

Now, to avoid both the error, radar designer has to make the compromise between these two types of error. A relatively high threshold will reduce reliability of false alarms but simultaneously missed the target too. So for setting of the threshold, radar designer must take care about the nature of radar applications ground environment and interferences created by the ground clutter, high hills, sea reflection etc., alongwith other relative parameters.

5.6. RADAR DETECTION AS HYPOTHESIS TESTING

There are two hypothesis for any radar measurement that is to be tested for the presence of a target. Any one of two hypothesis can be assumed to be true.

(*i*) The radar measurement is due to interference only, known as null hypothesis (H_0).

(*ii*) The radar measurement is the combined result of interference alongwith the target echoes is known as H_1.

The detection logic, examine each radar measurement to be tested and the best suited hypothesis is accounting for the measurement. Now if null hypothesis H_0 best accounts for the data, then system declares that a target was not present at the range, angle or doppler coordinates of the measurement and if H_1 best suited for the data, the radar system declares that a target was present.

The decision between the two hypothesis is an exercises in statistical decision theory. The analysis start with a statistical description of the probability density function that defines the measurement to be tested under each of the hypothesis.

The followings two PDF files are used, if the simple to tested is dedicated as Y.

$$Py(y/H_0) = \text{PDF of } y \text{ given that a target was not presents}$$

$$Py(y/H_1) = \text{PDF of } y \text{ given that a target was present.}$$

The analysis of radar performance is dependent on estimating there PDF's for the system. Generally the detection is based on N samples of data y_n forming a column vector **y**

$$\mathbf{y} = [y_0, ..., y_{N-1}]^t \qquad ...(5.1)$$

$Py(y/H_0)$ and $Py(y/H_1)$ are used, there are some probabilities of interest may be defined as

(*i*) **PD (probability of detection) :** The probability that a target is declared, H_1 is chosen that means a target is in fact present.

(*ii*) **PFA(Probability false alarm) :** The probability that a target is declared H_1 is choosen, when a target is in fact not present.

(*iii*) **PM (Probability of Miss) :** The probability that target is not declared, H_0 is chosen, when a target is in fact present two types of errors can be made in the declaration of the presence or absence of a target.

As illustrated in Fig. 5.1,

- Target is at sent, detector says target is present *i.e.* false alarm.
- Target is present, detector says target is absent *i.e.* missed detection.

Assume that we know the probability distribution of the measurement of received voltage $V \geq 0$ for the cases of (*i*), target absent and (*ii*) target is present. The case (*i*) is noise only and the latter is a signal voltage pulses. The noise Fig. 5.2 shows the two probability density functions such as $P_0(V)$ and $P_1(V)$. The probability that the measured voltage is between V and $V + dV$ is $P_i(V)\, dV$, where $i = 0$ represent target absent or $i = 1$, represents target is present.

The integral of each P_i from 0 to infinity is unity. Each P_i is characterized by a mean voltage μ_i.

The detection decision is made by choosing a threshold voltage, V_T and declaring.

- $V > V_T$ indicate target is present
- $V < V_T$ indicates target is absent.

The probability of false alarm (PFA) and the probability of correct detection (PD) are the integrals of P_0 and P_1 respectively from V_T to infinity.

In the presence of noise, it is not possible to achieve the ideal case of $P_D = 1$, PFA = 0. As V_T increases, PFA decreases but so does *PD* also. A curve of PD vs PFA (as V_T varies) is called a receiver operating characteristics (ROC) curve shown in Fig. 5.3.

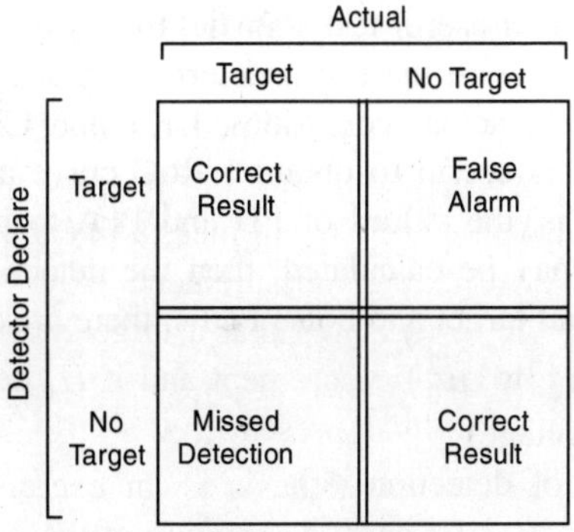

Fig. 5.3. Types of Target Declaration Errors.

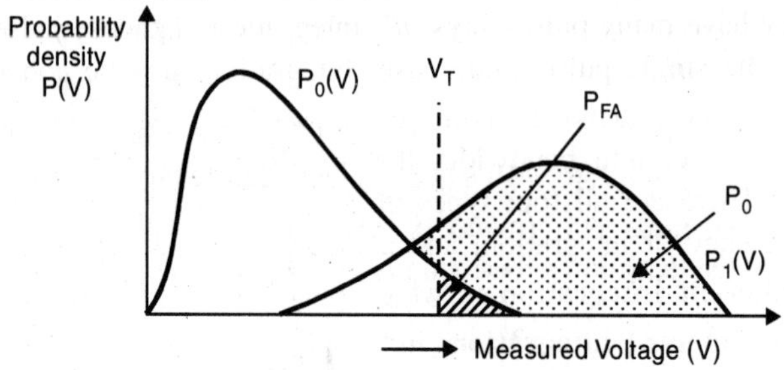

Fig. 5.4. Calculation of PD and PFA.

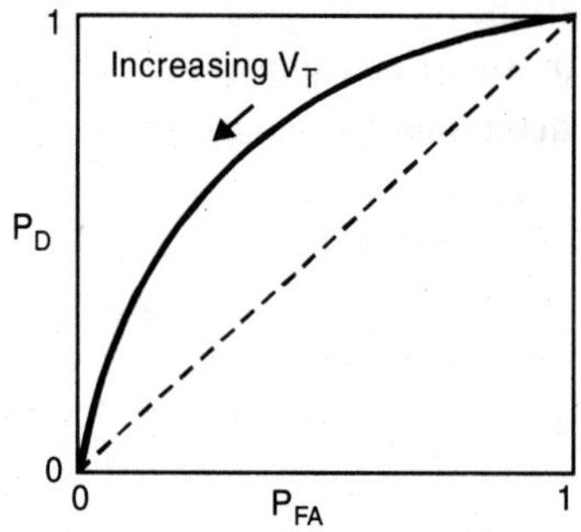

Fig. 5.5. PDC Curve.

The closer the curve is to the upper left corner p(PD = 1), (PFA = 0)], the better is the detection procedure.

The threshold V_T can be determined if we know the costs of false alarms (type-1 error) and missed detection (type-II errors). We describes

(*i*) CMD = Cost of missed detection = Cost of deciding target is absent (hypothesis H_0) if target is present (hypothesis H_1)

(*ii*) CFA = Cost of false alarm = Cost of deciding H_1 if H_0 is true.

(*iii*) Risk = $C_{MD}(1-P_D)N_T = C_{FA}\,P_{FA}N_A$;

where *NT* and *NA* are the priori numbers of opportunities for detections and false alarms, respectively.

The Bays criterion states that we should minimize the risk. We may define the ratio between $P_1(V)$ and $P_0(V)$ as the likelihood ratio $\Lambda(V)$. The Bays criterion than leads to the condition that

$$\Lambda(V) = \frac{P_1(V)}{P_0(V)} \underset{H_0}{\overset{H_0}{\gtrless}} \frac{C_{FA}N_A}{C_{MD}N_T} \qquad ...(5.1)$$

The **likelihood ratio** is a useful test statistic, that is a positive scalar quantity to be compared with a threshold, with the result determining whatever the target is declared to be present or absent. In actual conditions, CFA and CMD may be impossible to quantity, in such cases it is useful to obtain a RoC curve and allow the user to select the operating point, that is, the values of PD and PFA to be chosen.

If $P_0(V)$ and $P_1(V)$ may be calculated, then the detection problem can be solved. For given assumption about target and noise PDFs, there are four parameters of interest :

(*i*) SNR (ratio of μ_1^2 to μ_0^2)

(*ii*) *VT* (threshold voltage)

(*iii*) *PD* (Probability of detection)

(*iv*) PFA (Probability of false alarm)

We may have many pulses says 'n'; integration of thoese pulse can lead to a higher PD than in the single pulse value case. Integration may be coherent and non-current.

The detecability factor, D, is in pulsed radar, the ratio of single-pulse signal energy to noise power per unit bandwidth that provides stated probabilities of detection are false alarm. There are all four types of detections. The methods are listed in order of decling efficiency of integration, and also declining compresity of impletation.

(*i*) **Coherent Integration** in which pulses are added prior to envelope detection;

(*ii*) **Noncoherent integration,** in which each pulse is envelope detected and the resulting video pulses and are added together prior to application of thresholding.

(*iii*) **Binary integration,** in which each pulse is applied to a threshold, are the number M of threshold crossings is used as the criterion for an output alarm.

(*iv*) **Commulative detection,** in which $M = 1$, is the alarm criterion.

5.7. DETECTION THEORY

5.7.1. Neyman Pearson Observer

According to Neyamn Pearson observer theory the threshold level should be so choosen to have minimum type II error *i.e.* missed detection and fix the type I error.

Neyman Pearson have considered that this test in statistics for determining the validating the statistical hypothesis. In statistical term it is most powerful theory, and is on optimum one, no matter what the probability of signal and noise. Some times this type of threshold detection is known as Neyman Pearson observer. This criterion is most suitable for radar application.

5.7.2. Ideal Observer

"Siegert" that is known as ideal observer formulated one of the first mathematical criteria applied to radar detection theory. This criterion maximizes the ideal probability of correct decision or it minimizes the total probability of an error.

Both the types of statistical detection theory errors are considered by the ideal observer criteria and describe the probability of an error, which is given as

$$P(E) = P(N)\alpha + P(SN)\beta \qquad ...(5.11)$$

Where $P(N)$ and $P(SN)$ —— a prior probability of obtaining noise and signal along with noise

$$P(N) + P(SN) = 1$$

α = Conditional probability of false alarm given that noise is preset

β = Conditional probability of a miss

The prior probabilities are the probability of getting signal along with noise or noise alone before the detection is made. The probabilities must be known before applying this theorem.

5.7.3. Sequential Observer

In the above-mentioned theories, it is observed that the numbers of observations are fixed. However the detection decision can be made with the few observation or sometime with a single observation. The decision of detection taken by using a single observation only is called as sequential observer. In the sequential observer; a single observation can be taken in to account and is decided between one of three choices:

1. The R_x o/p is due to presence of signal along with the noise
2. The R_x o/p is due to presence of noise only
3. If the evidence is not adequate to take the decision between choices 1 and 2.

In the sequential observer, radar designer has two-threshold level, one is relatively lower than the other. If the R_x o/p is below than the lower threshold than it is taken as noise if the R_x o/p is enough to cross the upper threshold level than this is said to be the signal with the noise. But if R_x o/p is in between the two-threshold level, that means R_x o/p is sufficient enough to cross the lower threshold but not enough to cross the upper threshold level. In this case another observation is to be made to take the decision, and the process is repeated until the evidence is convincing enough to make a definite decision.

In the detection criteria, sequential observer theory is much more effective on the average than the Neyman- Pearson observer and ideal observer in which the test terminates after the (n) observation. This technique is very effective decision making criteria and can result in considerable amount of saving.

Sequential testing technique has been widely used in radar system for Quality control.

5.7.4. Likelihood Ratio

The above-mentioned detection criteria may be explained in terms of likelihood ratio. The likelihood ratio can be defined as the ratio of probability density factor of the signal plus noise, to the probability density function noise alone.

$$Lr(r) = P_{sn}(r)/P_n(v) \qquad ...(5.12)$$

Where $L_r(v)$ = likelihood ratio

$P_{sn}(r)$ = probability density function of the noise alone.

If the likelihood ratio is large enough, it may be treated, as signal is present.

The neyman Pearson observer is equivalent to examining the likelihood ratio and estimating

If $L_r(v) \geq K$

Where K is real positive number depending on the probability false alarm selected. The ideal observer criteria requires that the likelihood ratio $[L_r(v)]$

$$L_r(v) \geq P(n)/P(sn)$$

Where $P(n)$ and $P(sn)$ are the prior probabilities of the nth stage likelihood ratio $L_r(v)$ is compared with two threshold such that the signal is present if $Lr(v) = k_1$ and is absent if $L_r(vn) > k_2$ here $k_1$and k_2 are determined from probability of missing of detection and probability of false alarm.

If $k_2 < L_r(vn) < k_1$ the evidence is not enough and more observations are to be made.

5.8. DETECTOR CHARACTERISTICS

The portion of the radar receiver from the output of IF Amplifier to the input of radar display or data processor, which extracts the signal from the modulated carrier, is called the detector.

A detectors characteristic effects the detection of desired signal and the noise in the receiver. A good radar detector should consider the characteristics of the detector, which are probability of detection as a function probability of false alarm number of hits integrated and S\N ratio. There are basically, four types of detector used in the radar receiver.

1. Envelope detector or non-coherent detector
2. Logarithm detector
3. Zero crossing detector
4. Coherent detectors

Brief explanation of each detector is discussed in the present section.

5.8.1. The Envelope Detector : Non-Coherent Detector

The envelope detector is basically performing the function to extract the modulating signal from the carrier wave. This detection is purely based on the amplitude of the envelope as envelope detector and destroy all the information of phases.

In the initial stage of envelope detector a rectifying element is used that rectify the modulated signal and further passed through the low pass filler, which removed the higher frequencies (carrier frequencies). The output signal to the input signal of the rectifier is known as **detector law**.

There are two types of detector law characteristics, which are linear and square law characteristics. In the linear detector the output signal follows the input envelope linearly. So it is called **linear detector**. But detectors are not linear devices, they are non-linear devices. When the output signals squared of the input signal then this is called the **squared law detector**. In this detector the output, which we get from the detector, is directly proportional to the square of the input envelope. The difference between linear law and square law detector is very small. In radar when we discuss about the detector law then this include the video integrated also.

5.8.2. Logarithmic Detector

In this type of detector the O\P signal of the detector is proportional to the logarithm of the input signal and it is called a **logarithmic detector**.

This detector is used where more dynamic range is required and to avoid the saturation of the receiver from the reflections of the unwanted signals such as clutter. Generally ground reflection receiver from the nearly range, saturated the radar receiver, so to avoid the situation a logarithmic detector may be used.

5.8.3. Zero Crossing Detectors

This type of detector used for detection of signal and noise by virtue of the frequency within the signal crosses the zero level axis. The particular property of this detector is determine the distance between the crossings of the wave from along the zero level axis, may be related to the frequency of the waveform. In general, the greater S\N ratio, the smaller will be the number of zero crossing in a unit interval of time. If suitable device such as frequency counter is used to measure the average number of zero crossing in a unit interval time, a decision can be taken by counting the number of crossing in a unit interval time. If this number is less than a predetermined value, than it is said to be the presence of the target but when this number is more than a predetermined value, than it is said to be noise.

The Fig. 5.6(*a*) and (*b*) shown an example of detection by using zero detection criterion. In Fig. 5.6(*a*) the signal, consider a sine wave crosses the zero level six times and in Fig. 5.6(*b*) is crosses 8 times. The signal shown in Fig. 5.6(*a*) is a target signal but the signal shown in Fig. 5.6(*b*) is the noise; the average number of zero crossing might be increased.

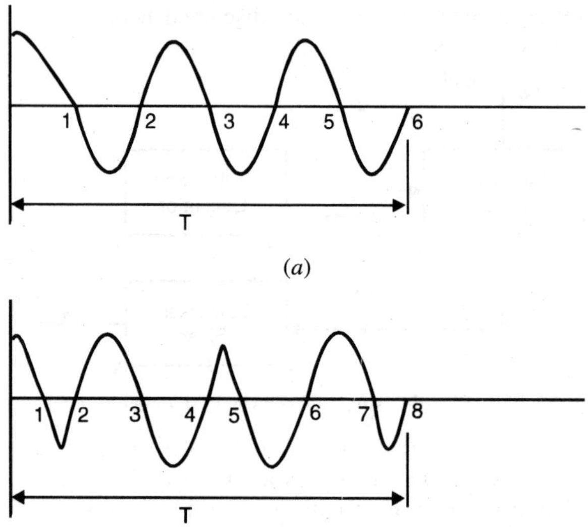

(*a*)

(*b*)

Fig. 5.6 (*a*) and (*b*). (*a*) Sine Wave with 6 Zero Crossings in Time *T*, (*b*) Sine Wave Noise with 8 Zero Crossing in Time *T*.

According to the Gaussian, the average number of zero crossing per second at the O\P of narrow band pass filter of rectangular shape when the input is a sine wave in Gaussian noise given by the following equations.

$$\overline{n_0} = 2f_0\{S\backslash N + 1 + (f_B^2/12f_0^2)/(S/N + 1)\} \quad ...(5.13)$$

Where $\overline{n_0}$ = Average number of zero level axis crossing.

f_0 = Center frequency of filter

f_B = signal to noise (power) ratio.

These zero crossing information used in the period meter, in this the periodically each cycle is measured and displayed on the CRO as the height of a spot above the base line and other method is the phase filter, in this method the O\P of the phase filter is dependent of its amplitude.

5.8.4. Coherent Detector

The coherent detector converts the high frequency carrier in to low frequency or dc components. It does not extract the information from the carrier. It does not loose the information of phase or amplitude as in the detector discussed earlier. A simple block diagram of coherent detector is shown as Fig. 5.6.

This configuration uses three devices to perform the coherent detection. The input to the balanced mixer is a signal of known frequency (f_o) and known phase (Q_o) along with the noise. The reference oscillator is being used to provide the signal as the input signal with the same frequency and phase to be detected. The balanced mixer receive both the input from the reference oscillator and the input signals that is to be detected. The output of balanced mixer is pass through the low pass filter that allows only the DC or low frequency modulation components and it rejects the high frequencies carrier wave.

This detector is a truly linear detector, whereas a linear envelop detector, which had been discussed earlier, is not a linear in the same sense. So therefore, the coherent detector is more efficient than other detector discussed here.

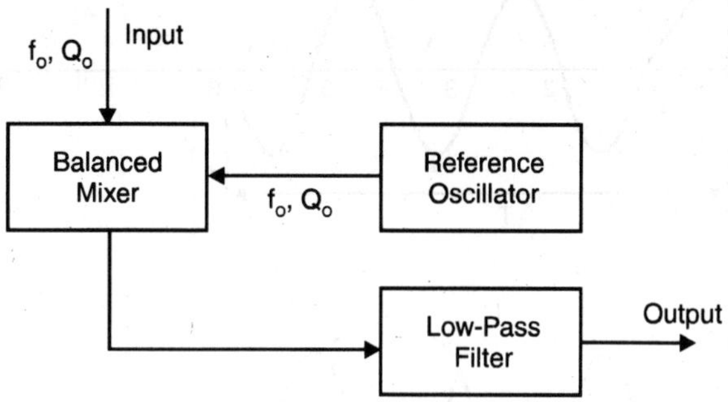

Fig. 5.7. A Simple Block Diagram of Coherent Detector.

The S/N ratio is much better in coherent detector than the linear detector and the zero crossing detectors. The coherent detector improves the S/N ratio from 1 dB to 3 dB or more. This detector is quite similar to the cross correlation receiver.

5.8.5. Synchronous Detection using Phase Locked Loop

In the synchronous detector the phase of the reference signal is synchronized with the incoming signal. By using some device, the phase of the reference signal is synchronized with the incoming signal. This technique is known as synchronous detector. The device used to maintain the phase of the reference oscillator synchronized with the incoming signal, is known as Phase locked loop. The Fig. 5.8 shows the simple block diagram of phase locked loop configuration. In this technique servo control loop is used to maintain the phase of the voltage controlled reference oscillator in step with the phase of incoming signal. However the input S/N ratio should not be so low otherwise reference signal oscillator will synchronize the noise instead of signal.

Synchronous detection is quite difficult in pulse system but it is easier to detect the signal of continuous nature.

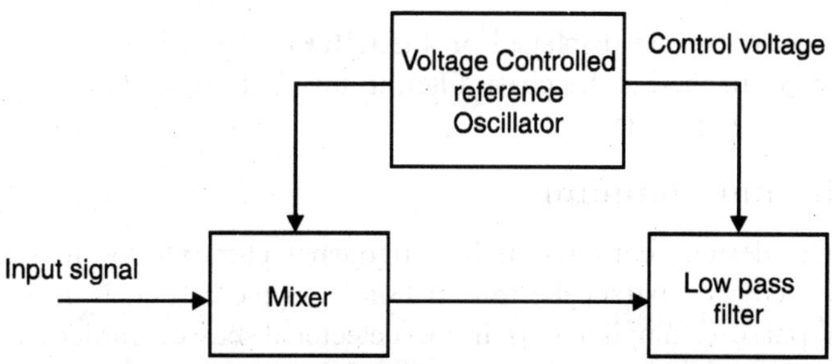

Fig. 5.8. Synchronous Detection using Phase-Locked-Loop.

5.8.6. I & Q Demodulator

Synchronous demodulation has three problem, first it does nto recover enough information about the signal to tell whether the Doppler shift is positive or negative. Second, only those components of the signal which are in phase with the COHO are recovered, those at 90° phase (the quadrature components) are lost. This results in a reduction in signal-to-noise ratio of 3 db from that which would be available if all components of the signal were recovered. Third signal of certain phases, known as blind phases, are lost in synchronous demodulation. These are signals whose phase with respect to the COHO are ± 90°, the cosine of which are zero. Some other use the term synchronous demodulation synoymously with "I/Q demodulator".

An I/Q demodulator, Fig. 5.9 recover all components of the signal, gives sufficient information to discriminate between positive and negative doppler, and has no blind phases, it is two synchronous demodulators, one comparing the signal to the COHO and the other comparing the signal to the COHO shifted 90°. The signals recovered are given in the equation below.

$$V_I = V_s \cos [\phi t]$$
$$V_Q = V_s \sin [\phi t]$$

where, $V_I = V_{\text{out}}$ of I demodulator

$V_Q = V_{\text{out}}$ of ϕ demodulator

V_s = The signal peak voltage

ϕt = The signal phase as a function of time, rerefers the COHO.

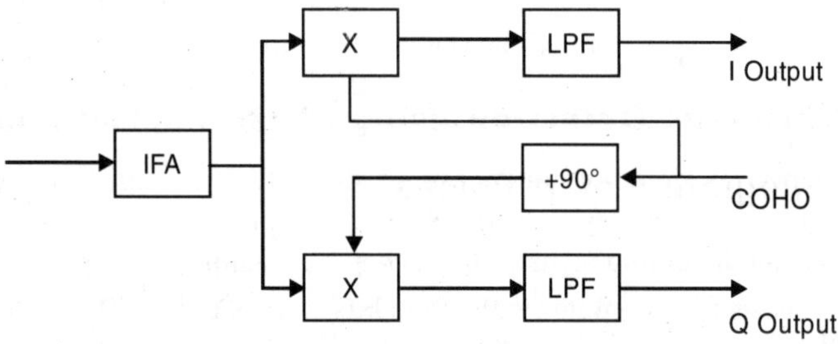

Fig. 5.9. I/Q Demodulator.

5.9. DETECTION OF NON-FLUCTUATING TARGET IN NOISE

We will consider detection of a point target referred as non-fluctuating.

Coherent detection of a single pulse of known phase

$$\text{DC}(1) = [erfc^{-1}(2P_{FA}) - erfc^{-1}(2P_D)]^2 \quad \text{...(5.14)}$$

where we use a function derived from the error function $erf(v)$,

$$\left.\begin{aligned} erf(v) &\equiv \frac{2}{\sqrt{\pi}} \int_0^V \exp(-v^2)\,dv = X,\ erf^{-1}(X) = V \\ erfc(v) &\equiv 1 - erf(v) = 1 - X,\ erfc^{-1}(1-X) = V \end{aligned}\right\} \quad \text{...(5.15)}$$

Envelope or non-coherent detection of a single pulse

According to the North's approximation,

$$D_0(1) \approx \left[\sqrt{In\left(\frac{1}{P_{FA}}\right)} - erfc^{-1}(2P_D)^2 - \frac{1}{2}\right] \quad \text{...(5.16)}$$

It is instructive to examine this important case, for a single pulse, we denote the following outputs from envelope detector :

r = Magnitude of noise voltage (variable)

A = Magnitude of signal (target) voltage (constant)

According to the rice for a squared-law detector and summarized the PDF of r, the received voltage for the target present case is

$$P_1(r) = \frac{r}{\sigma_v^2} \exp\left[-\frac{(r^2 + A^2)}{2\sigma_v^2} . I_0\left(\frac{rA}{\sigma_v^2}\right)\right] \quad ...(5.17)$$

where A is the received signal (without noise) voltage, $I_0(z) = I_0(j2)$, is a modified Bessel function. This is a Rician PDF. For the target absent case, we get $A = 0$ and obtain the PDF in noise only :

$$P_0(r) = \frac{r}{\sigma_v^2} \exp\left(-\frac{r^2}{2\sigma_v^2}\right) \quad ...(5.18)$$

$$\text{PFA} = \int_{VT}^{\alpha} \frac{r}{\sigma_v^2} e^{-r^2/2\sigma_v^2} dr = e^{-v_T^2/2\sigma_v^2} \quad ...(5.19)$$

A good approximation to PD is

$$\text{PD} \cong \frac{1}{2}\left[1 - erfc\left(\sqrt{In\left(\frac{1}{P_{FA}}\right)} - \sqrt{SNR + \frac{1}{2}}\right)\right] \quad ...(5.20)$$

Coherent Integration of 'n' Pulses :

When 'n' pulses are coherently integrated, the resulting SNR is n times the SNR for a single coherently integrated pulse as shown in Fig. 5.4. Thus,

$$\frac{V_s}{|V_N|} = n^{1/2} \frac{V_s}{(V_n^2)^{1/2}}$$

$$\text{SNR} = \left(\frac{V_s}{|V_N|}\right) = n\,(\text{SNR, single phase})$$

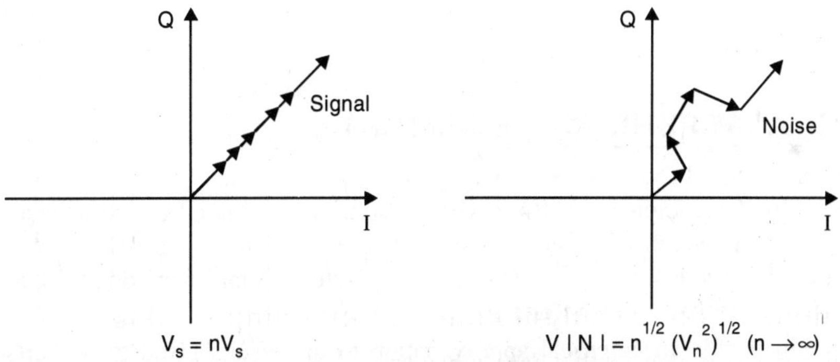

Fig. 5.10. Coherent Integration.

$$D_C(n) = \frac{D_C(1)}{n}$$

Fluctuation Loss : Fluctuation loss is the required increase in signal energy for detection of a fluctuating target compared with a non-fluctuating target. For a swerling-I target, single pulse,

$$L_f(1) \equiv \frac{D_1(1)}{D_0(1)}$$

5.10. BINARY AND COMMULATIVE INTEGRATION : NON-FLUCTUATING TARGET

The terms **binary integration** refers to the case for which we gather information from a number of dweels, N, declare for each that the target is present or absent (a binary decision), then combine the results of the binary decisions. The following discussions applies to a non-fluctuating target, where the probabilities for successive dwells are independent; the case for a fluctuating target is more complicated. Thus

(*i*) The probability of detecting the target in one dwell is P_D.

(*ii*) The probability of not detecting the target in one dwell is $1 - P_D$.

(*iii*) The probability of not detecting the target is 'N' dwells is $(1 - P_D)^N$.

(*iv*) The probability of at least one detection in N dwells is $1 - (1 - P_D)^N$.

For 'M' detection in 'N' dwells.

(*i*) The probability of exactly M detection in M dwells is

$$P'_D(M, N) = \binom{N}{M} P_D^M (1 - P_D)^{N-M}$$

where

$$\binom{N}{M} = \frac{N!}{(N-M)!\,M!} \qquad ...(5.21)$$

(*ii*) The probability of at least M detections in N dwells (M out of N detection) is

$$P_D(M, N) = \sum_{k=M}^{M} \binom{N}{k} P_D^k (1 - P_D)^{N-k} \qquad ...(5.22)$$

Binary integration is often performed for convenience, for example, for a relatively slow scanning radar searching for a target in non-stationary clutter. However, if the clutter distribution is stationary, the overall probability of detection is less for binary integration than for non-coherent integration.

Commulative detection is equivalent to binary detection with $M = 1$.

5.11. ALBERSHEIM'S EQUATION

A simple closed-form expression relating to P_D, P_{FA} and SNR, is given by the Albersheim in 1981, known as **Albersheim's equation.** In this all the parameters such as P_D, P_{FA} and SNR can be computed with the simple scientific calculators. It is seen that the performance result for the case of a non-fluctuating target in complex Gaussian noise required modern software analysis.

This equation is an empirical approximation to the results given by Robertson in 1967 for calculating the single-sample SNR χ_1, required to obtain a given P_D and P_{FA}. The following condition are required,

(*i*) Linear detector

(*ii*) Non-fluctuating target in Gaussian noise

(*iii*) Non-coherent integration of N samples

The single-sample SNR χ_1 may be estimated by the use of following relation

$$\chi_1 = -5\log_{10} N + \left[6.2 + \left(\frac{4.54}{\sqrt{N+0.44}}\right)\right]\log_{10}(A + 0.12\,AB + 1.7B)\text{ dB} \quad ...(5.23)$$

$$A = \ln\left(\frac{0.62}{P_{FA}}\right)$$

$$B = \ln\left(\frac{P_D}{1-P_D}\right)$$

where χ_1 is in dB's not in linear power units.

While estimating the χ_1, the error is less than 0.2 dB for $10^{-7} \le P_{FA} \le 10^{-3}$, $0.1 \le P_D \le 0.9$, and $1 \le N \le 8096$

If we take the case of $N = 1$, then

$$A = \ln\left(\frac{0.62}{P_{FA}}\right)$$

$$B = \ln\left(\frac{P_D}{1-P_D}\right)$$

$$\chi_1 = 10\log_{10}(A + 0.12\,AB + 1.7B)\text{ dB}$$

5.12. SHNIDMAN'S EQUATION

Albersheim's equation provided a simple approximation for the non-fluctuating target, but that estimation is not applicable to fluctuating targets. The Shnidman's equation given by Shnidman in 2002. It is similar to Albersheim's equation in this series of equations gives the single-pulse SNR χ as required to obtain a specified P_D and P_{FA} with non coherent integration of N samples. In this equation results are for a square law detector.

The following series of calculation will gives the Shnidman's equation.

$$K = \begin{cases} \infty \text{ for non-fluctuating target (swerling } 0/5) \\ 1 \text{ swerling 1} \\ \text{N swerling 2} \\ 2 \text{ swerling 3} \\ \text{2N swerling 4} \end{cases}$$

$$\mu = \begin{cases} 0 & N < 40 \\ \frac{1}{4} & N \ge 40 \end{cases}$$

$$\eta = \sqrt{-0.8\ln[4P_{FA}(1-P_{FA})]} + (P_D - 0.5)\sqrt{-0.8\ln[4P_D(1-P_D)]}$$

$$X_\infty = n\left[\eta + 2\sqrt{\frac{N}{2} + \left(\alpha - \frac{1}{4}\right)}\right]$$

$$C_1 = \{[(17.7006P_D - 18.4496)P_D + 14.5339]\,P_D - 3.525\}/K$$

$$C_2 = \frac{1}{k}\left\{\exp(27.31P_D - 25.14) + (P_D - 0.8)\left[0.7\ln\left(\frac{10^{-5}}{P_{FA}}\right) + \frac{(2N-20)}{80}\right]\right\}$$

$$C\text{ dB} = \begin{cases} C_1 & 0.1 \le P_D \le 0.872 \\ C_1 + C_2 & 0.872 < P_D \le 0.99 \end{cases}$$

$$C = 10^{\mathrm{C}}_{\mathrm{db}/10}$$

$$\chi_1 = \frac{C \cdot \chi_\infty}{N}$$

$$\chi_1\,(\text{dB}) = 10 \log_{10}(\chi_1)$$

The accuracy of Shnidman's equation is not as good as given by Albersheim's equation. Except at the extreme values of P_D for the swerling 1 model, the error in the estimate of χ_1 is less than 0.5 dB for $0.1 \le P_D \le 0.99$, $10^{-9} \le \text{PFA} \le 10^{-3}$ and $1 \le N \le 100$. Even though some error may be there in the result from Shnidman's equation but still it appears to be the most useful general tool presently available.

5.13. AUTOMATIC DETECTION

When the decision of detection operation is performed by electronically method automatically without intervention of radar operator, then this is called Automatic detection. Radar operator, who is efficient to detect the target within the noise, in some cases, may miss the target due to fatigue, characteristics of CRT, boredom etc. The automatic detection provides the detection decisions either to the radar operator for action or to a computer device for further processing. In some cases the radar operator is more capable of discriminating the target within the clutters and interference but when the number of target present on the scopes are more than in this situation, radar operator may not be able to detect the target efficiently. The automatic detection require much more efficient receiver to detect the targets.

The automatic detection perform the following steps while detecting the targets.

1. Quantisation. It means quantisation of total radar coverage, either in range, or may be in angle, resolutions cells.

2. Sampling. Sampling provides the samples for each range—resolution cell at least each sample for per resolution cell. Practically it should be more than one samples for each cell to have complete information.

3. A\D converter. A\D converter, converts the analog samples in to digital word, each samples is converted its corresponding digital value (In Binary).

4. Signal Processing. Some processing of the signals are being done by using some electronic devices to remove or eliminate the noise, interference clutter echoes before attempting the detections decision

5. Integration. It will integrate the available samples for each resolution cell.

6. Constant false alarm. CFAR circuit will maintain the false alarm rate constant.

7. Clutter map. Clutter mapper will maintain the received signals with the known clutter location and ignore the known clutter echoes.

8. Threshold detection. Threshold detection decides, either incoming signal from the target or from the noise. All those signal which are crossing the threshold level are made to pass through for further processing rest are blocked there itself and treated as noise.

Finally after the threshold detection, the detection decision is made; measurement of target and range angle can be given.

An automatic detection and tracking system has an advantage to handle many targets and determine if clutter realize tracking.

5.14. CONSTANT FALSE ALARM RATE (CFAR)

A constant false alarm rate is an electronic device, which maintains the false alarm rate constant by raising the threshold level automatically to block the clutter echoes and external interferences from overloading the automatic tracker. As it is already discussed earlier that a target is detected when the o/p of the radar receiver crosses a predetermined fixed threshold to achieve a probability of false alarm. In many condition when clutter echoes or some interference may be large enough than the receiver internal noise and they may cross the threshold level and false alarm may occur.

When at the earlier stage, automatic detection and tracking system was incorporated in the existing radar, which did not have the good clutter rejection facility, a need of constant false alarm rate (CFAR) was recognized. CFAR is being used, at the cost of a lower probability of detection of desired target.

5.15. CELL AVERAGING CFAR

In Fig. 5.11, a cell average CFAR has shown, which is mostly used in the radar receiver for the detection. A cell average CFAR does not sum all the range cells but sums the cells ahead of the test cell separately from the sum of the cells following the test cell. It uses an adaptic threshold, where level is decided by the clutter or noise in the vicinity of the radar echo. The o/p of the test cell is the radar video, which is compared to the variable threshold, derives from the some of the outputs of the tapped delay lines, which are generally 16 to 20 in numbers, defining the reference signal.

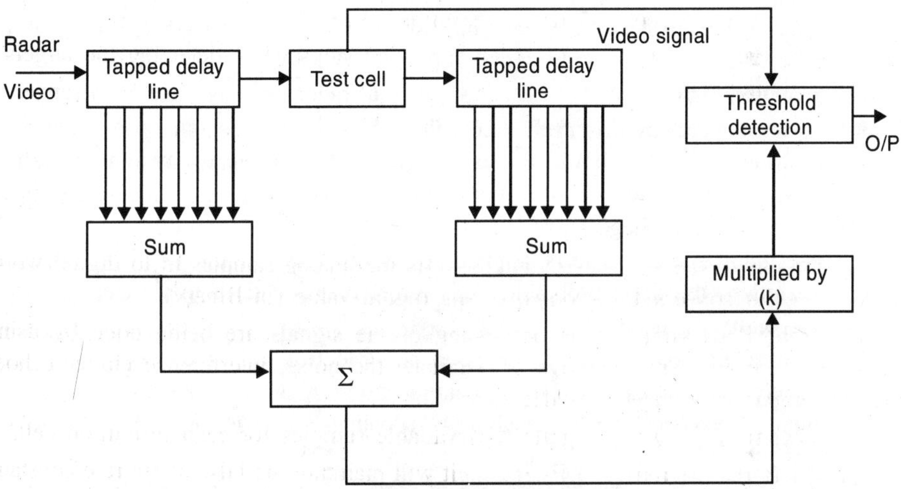

Fig. 5.11. A Block Diagram of Cell Averaging Constant False Alarm Rate.

SUMMARY

- The radar has to perform two basic functions such as detection of the signal in the noise and second is extraction of target information in the term of position, velocity and size of the target.
- The matched–filter receiver is an optimum method for detection of reflecting signals in noise. A characteristic of the radar receiver, which makes the frequency

response function maximize to the output peak signal-to mean-noise ratio, is called a **matched filter**. It is equal to the $2E_0/N_0$, where E_0 is the received signal energy and N_0 is noise power in the bandwidth.

- There is one thumb rule in pulse radar that the bandwidth of receiver should be approximately equal to the reciprocal of the pulse width (T), it can be written as

$$BW \approx 1/T$$

- **Correlation Detection.** The basis of this determination involves the computation of the correlation between the two signals, the outgoing and incoming. **Correlation** is a measure of the similarity or relatedness between two waveforms. For two waveforms v_1 and v_2 the correlation is mathematically given by the following equation

$$R(t) = \int_{-T/2}^{T/2} v_1(t')\, v_2(t+t')\, dt'$$

If $\quad T \to \infty$

- **Auto correlation** is the measure of the coherence of a waveform with itself. It is noted that in the case of auto correlation it would be maximum when the time shift would be zero or a multiple of its time period.
- **Correlation** is the **optimal** technique for detecting a known waveform in random noise. Using correlation to detect a known waveform is frequently called **matched filtering**.
- **Cross correlation** is measure of the coherence between two pulses. The principle of the cross correlation function may be used as a basis for radar receiver design.
- **Detection Criteria.** It is mean interest of radar engineer that the received signal is either from the radar or from the target. The detection of the weak signal in the presence of noise to decide whether the received signal is due to noise or noise along with signal.
- **Detection Theory**
- **Neyman Pearson observer.** According to Neyman Pearson observer theory the threshold level should be so chosen to have minimum type II error *i.e.* missed detection and fix the type I error.
- **Ideal observer.** Siegert that is known as ideal observer formulated one of the first mathematical criteria applied to radar detection theory. This criterion maximizes the ideal probability of correct decision or it minimizes the total probability of an error.
- **Sequential observer.** The decision of detection taken by using a single observation only is called as sequential observer. In the sequential observer; a single observation can be taken in to account and is decided between one of three choices:
    1. The R_x o/p is due to presence of signal along with the noise
    2. The R_x o/p is due to presence of noise only.
    3. If the evidence is not adequate to take the decision between choices 1 and 2.
- **Likelihood ratio.** The likelihood ratio can be defined as the ratio of probability density factor of the signal plus noise, to the probability density function noise alone.

$$L_r(r) = P_{sn}(r)/P_n(v)$$

where $L_r(v)$ = likelihood ratio

$P_{sn}(r)$ = probability density function of the noise alone.

If the likelihood ratio is large enough, it may be treated, as signal is present.

- The portion of the radar receiver from the output of IF Amplifier to the input of radar display or data processor, which extracts the signal from the modulated carrier, is called **the detector**.
- There are basically, four types of detector used in the radar receiver. Which are as follow:
  1. Envelope detector
  2. Logarithm detector
  3. Zero crossing detector
  4. Coherent detectors
- **The Envelope Detector.** The envelope detector is basically performing the function to extract the modulating signal from the carrier wave. This detection is purely based on the amplitude of the envelope as envelope detector and put losses all this information of phases.
- The output signal to the input signal of the rectifier is know as **detector law**.
- In the linear detector the output signal follows the input envelope linearly. So it is called **linear detector**.
- When the output signals squared of the input signal then this is called the **squared law detector**.
- **Logarithmic Detector.** In this type of detector the O/P signal of the detector is proportional to the logarithm of the input signal and it is called a **logarithmic detector**.
- **Zero crossing detectors.** This type of detector used for detection of signal and noise by virtue of the frequency within the signal crosses the zero level axis. The particular property of this detector is determine the distance between the crossings of the wave from along the zero level axis, may be related to the frequency of the waveform.
- **Coherent detector.** The coherent detector converts the high frequency carrier to be low frequency or dc components. It does not extract the information from the carrier. It does not loose the information of phase or amplitude as in the detector.
- **Synchronous detection using phase locked loop.** In the synchronous detector the phase of the reference signal is synchronized with the incoming signal. By using some device, the phase of the reference signal is synchronized with the incoming signal. This technique is known as synchronous detector. The device used to maintain the phase of the reference oscillator synchronized with the incoming signal, is known as **Phase locked loop.**
- **Automatic detection.** When the decision of detection operation is performed by electronically method automatically without intervention of radar operator, then this is called **Automatic detection**.
- **Constant false Alarm Rate (CFAR).** A constant false alarm rate is an electronic device, which maintains the false alarm rate constant by raising the threshold level automatically to block the clutter echoes and external interferences from overloading the automatic tracker.
- There are two hypothesis for any radar measurement that is to be tested for the presence of a target. Any one of two hypothesis can be assumed to be true.
  (*i*) The radar measurement is due to interference only, known as null hypothesis (H_0).
  (*ii*) The radar measurement is the combined result of interference alongwith the target echoes is known as H_1.

- There are some probabilities of interest may be defined as
  - (*i*) **PD (Probability of detection) :** The probability that a target is declared, H_1 is chosen that means a target is in fact present.
  - (*ii*) **PFA (Probability false alarm) :** The probability that a target is declared H_1 is choosen, when a target is in fact not present.
  - (*iii*) **PM (Probability of miss) :** The probability that target is not declared, H_0 is chosen, when a target is in fact present two types of errors can be made in the declaration of the presence or absence of a target.
- **Coherent Integration** in which pulses are added prior to envelope detection;
- **Noncoherent integration,** in which each pulse is envelope detected and the resulting video pulses and are added together prior to application of thresholding.
- **Binary integration,** in which each pulse is applied to a threshold, are the number M of threshold crossings is used as the criterion for an output alarm.
- **Commulative detection,** in which $M = 1$, is the alarm criterion.
- **Fluctuation loss** is the required increase in signal energy for detection of a fluctuating target compared with a non-fluctuating target.
- A simple closed-form expression relating to P_D, P_{FA} and SNR, is given by the Albersheim in 1981, known as **Albersheim's equation.**
- The Shnidman's equation given by Shnidman in 2002.
- **Clutter** is defined as an unwanted radar echoes.
- Clutter can be due to reflection from storm clouds, clean air tubulance, precipitation and other meteorological phenomena, may confuse the radar operator.
- Generally radar clutter are being classified as of two types :
  - (*i*) Ground clutter or surface clutter
  - (*ii*) Sea clutter.
- The reflection of radar energy from the surface of the sea is known as **sea echo** or **sea clutters.**
- Clutter reductions techniques are to be used to minimise the effect of clutter in the radar.
- Sometime, radar echoes can be received from the regions of the atmosphere where no apparent reflecting sources exist. These echoes are known as "phantonis", "ghosts", "pixies" and "angels". The most popular name for these type of radar echoes is "Angels".

REVIEW QUESTIONS

1. What do you understand by detection of signal in noise ?
2. What is matched filter in radar receiver and write the equation for the North filter ?
3. What is the effect of S/N ratio over the detection of weak signals ?
4. What do you understand by correlation detection ?
5. Explain cross correlation with the help of Block diagram.
6. What are the different detection therories ? Explain them in brief.
7. What are be difference types of detection used in rader ? Explain zero crossing detector with the diagram.
8. What do you understand by automatic detection and explain the steps involved in this process ?
9. Explain cell averaging CFAR with block diagram.

10. Draw Constant False Alarm Rate (CFAR) radar receiver. Explain its working. **[UPTU, 2003]**

11. Obtain derivations of matched-filter characteristics of receiver. **[UPTU, 2003]**

12. Define radar clutter and explain how they affect the performance of the radar.

13. Explain terms radar clutter and explain various types of radar clutters.

14. Briefly discuss about the clutter cross-sections.

15. Briefly discuss the ground clutter and its affect over the radar.

16. Briefly discuss about the sea clutter.

17. What do you understand by angels ?

18. Write shorts notes :

(*a*) Auto correlation (*b*) Detection criteria
(*c*) Neyman Pearson observer (*d*) Ideal observer
(*e*) Sequential observer (*f*) Likelihood ratio
(*g*) Envelope detector (*h*) Logarithm detector
(*i*) Coherent detector (*j*) CFAR
(*k*) Clutter (*l*) Land clutter
(*m*) Sea clutter (*n*) Angels
(*o*) Antenna beam (*p*) Binary integration

19. Explain the detection criteria in radar signals.

20. Define threshold detections in radar signals.

21. Briefly discuss about the radar detections as hypothesis testing.

22. Explain the detections of Non-fluctuating target in noise.

23. Give the Albersheim's equations and explain the accuracy of this equation.

24. Give the Shnidman's equation and explain the error in this equation.

❑❑❑

CHAPTER

6

INFORMATION FROM RADAR SIGNALS

INSIDE THIS CHAPTER

6.1. BASIC RADAR MEASUREMENTS

The observation of radar signals in the presence of noise may be divided in two sections for the purpose of analysis :

(*i*) The detection of the signal in the presence of noise.

(*ii*) The extraction of target informations available in the radar signals.

About the detection of the signal, we have already discussed earlier in Chapter 10. In this unit, we will consider several topics related to the extraction of informations from the target signals.

Some characteristics of the target that might be extracted from the radar signals. These are as follows :

(*i*) Range of the target

(*ii*) Doppler velocity

(*iii*) Angle of elevation and azimuth

(*iv*) Size of the target

(*v*) Shape of the target, and

(*vi*) Change of shape of the target.

As for the purpose of detection or recognition of the presence of a signal in noise is related to the statistical hypothesis. Many of the tools of mathematical statistics are applicable to the analysis of radar detection. Similarly, the extraction of informations from radar signals is also a problem of the statistical estimation of parameters.

The process of detection of signal and the extraction of information from the signal, is not totally independent. They both are meaningless without the other.

In this chapter, some of the fundamental concepts involves in the extraction of information from a radar signal are discussed. In the further sections, we will be discussing about the techniques used to extract the information from the radar signals.

6.2. MEASUREMENTS OF PHASE AND AMPLITUDE

This section considers in a qualititive manner the type of target information that may be obtained from measurements of the phase and amplitude of the echo signal as a function of direction, frequency and time. The source of target information in a radar system is nothing but the electromagnetic radiations scattered by the target in the direction of the radar receiver. The scattered signal in the space may be resolved into frequency spectrum at any point and at any given time each frequency components can be considered as having and amplitude and a phase. The radar measures the amplitude and the phase of the echo signal received at a given position at a given frequency and time. A single observation of phase and amplitude at a particular time, frequency and positions given some informations related to the target. However, information related to the target can be obtained by observing a series of amplitude and phase measurement as a function of time, frequency and position.

There are six possible first derivatives from which the nature of the target can be analysed. The measurements of angular position (direction), range and Doppler velocity may be derived from the variation of phase with position, frequency and time.

Measurements of target shape, size and rotation may be obtained by measuring the variation of the amplitude with the same dimensions. Depending upon the specified application of radar, the radar utilize the information available from the entire six derivatives. Mostly radars extracts only the phase variations.

The measurement of amplitude and phase as a function of position, frequency and time may be taken continuously or at discrete sampling intervals. Sampled measurements are easy and simple to implement but they can lead to ambiguities. When measurements are continuous than there is no ambiguities.

6.2.1. Phase Derivatives

Usually, the variations of phase with position, time and frequency are the measurements made by radar.

Observation of the phase of the received signals at the antenna as a function of position, when time and frequency remain constant, determine *the angular direction of the target.* The greater the phase difference per unit solid angle, more will be the accuracy in the measurement. If the spacing between the antenna elements is less than the $\frac{\lambda}{2}$ then there is no ambiguity in angle would result.

The change of phase with time, when position and frequency are fixed, is essentially the measurement of the Doppler frequency shift. It is assumed that the illuminating radiation is monochromatic and that the transmitted signal is used as reference. As in the case of CW radar, if the measurement of phase is continuous the Doppler measurement is unambiguous. In case of MTI or the pulse Doppler radar. When the measurement is not continuous, ambiguities may occur.

The variation of phase with frequency, when time and position remain constant then it is a measure of *the range of the target.* The rate of change of phase with frequency is equivalent to time delay or range. This principle is used in frequency CW radar, where the range to a target is determined by measuring the phase difference between two echo signals which are at different frequencies. The radars which transmits a pulse train and the frequency CW radar measure range discontinuously, so that ambiguities are possible. A continuous range measurement can be made with a frequency modulated transmission increasing linearly with time without ambiguities.

6.2.2. Amplitude Derivatives

The variations of signal amplitude as a function of frequency, assuming time and position remain constant, is a measure of the *size of the target.* In general the variations of amplitude with frequency is related to the size of the target.

The variations of signal amplitude with position, when the frequency and time are constant, is related to the *shape of the target.* For example, the relative charge of position of aircraft in close formation, the change of altitude of a single aircraft, result in a change of amplitude with time.

6.2.3. The Six Measurement Derivatives

The measurement of the change in amplitude with position x, frequency f, and time t, are not used as often as the corresponding phase measurements. Modern radars which are to be used to the fullest extent of their capabilities requires control or knowledge of the phase of the transmitted signal.

The six measurement derivatives are as follow :

1. $\left(\frac{\partial\phi}{\partial x}\right)_{t,f}$ $\longrightarrow$ Angle

2. $\left(\frac{\partial\phi}{\partial f}\right)_{t,x}$ $\longrightarrow$ Range

3. $\left(\frac{\partial\phi}{\partial t}\right)_{x,f}$ $\longrightarrow$ Relative velocity

4. $\left(\frac{\partial A}{\partial x}\right)_{tf}$ $\longrightarrow$ Shape

5. $\left(\frac{\partial A}{\partial t}\right)_{x,f}$ $\longrightarrow$ Change of shape

6. $\left(\frac{\partial A}{\partial f}\right)_{t,x}$ $\longrightarrow$ Size

6.3. REVIEW OF RADAR MEASUREMENTS

Most often, the three measurements made by radar are as follow :

(*i*) The range to the target

(*ii*) The rate of change of range

(*iii*) The angle of arrival

A radar can obtain a target location in range, azimuth and elevation. The target trajectory or track can be obtained, after several observation of a moving target over a period of time.

Range : This is the most important measurement given by the radar. The measurement of range is the measurement of the time difference between the arrival of the new signal and a reference signal. The reference is generally the transmitted signal.

The transmitted signal should be marked in such a manner that its echo may be identified as having been transmitted at a particular line known at the receiver. The marking of the transmitted pulse in the may be applied to the signal amplitude (pulse radar), frequency in FMCW, radar or phase in multiple-frequency CW radar.

The most common method of marking the transmitted signal is with amplitude modulation in the form of rectangular pulse. The transmitted signal may also be marked in time by varying the frequency which is used in FMCW radar. The frequency modulation may be carried out continuously with time or it may modulate a single long pulse as in the pulse compression technique. The phase of the transmitted signal is a time reference. It may be used to measure range by comparing the phase of the echo signals and the transmitted signal. Along pulse can be phase modulated to obtain a range measurement of high accuracy.

Range Gate : The change of range with respect to time may be determined from the shift in echo signal frequency caused by Doppler effect. The MTI radar, CW radar and pulse Doppler radar depend on the Doppler shift for discriminating stationary from the moving target. Range gate may be found by measuring the rate of change of successive observation of range.

Angular Position : The angular position of a target is specified by two orthogonal angular co-ordinates as azimuth and elevation. The angular coordinates of a target are estimated from the direction of the equiphase surface of the echo wavefront. This direction is called the angle of arrival.

The radar measurement of angle using a mechanically scanned parabolic antenna with single horn feed is essentially continuous and hence it is unambiguous. The larger the antenna, the narrower will be the antenna beam and the more accurate will be the measurement of angular positions. A highly directive antenna—one with narrow beam is scanned past the target and the direction corresponding to maximum echo signal amplitude indicates the direction of the target.

Angular accuracy may be improved at the cost of ambiguities by using two or more antenna located many wavelength apart. This is an example of a discontinuous angle measurement and is known as **interferometer.** The larger the spacing between the separate antennas, the more accuracy in angle measurement.

Measurement of Height

In many of the radar application, we don't need the measurement of the height. However there are many application in which knowledge of the target's position in three dimensions needed and this types of radar is known as **3-D radar.** The elevation angle can be used as the third coordinate but it is often more convenient to use height. Height can be determining from the measurement of range and elevation angle. Use of height is more suitable aircraft targets or for satellites with nearly circular orbits.

Three dimensional position information can be obtained with a symmetrical pencil beam antenna. The azimuth and elevation angle, can be determined from a single observations with a single radar set. The pencil beam may reason a hemispherical volume in space by rapidly nodding in elevation and slowly rotating in azimuth, or alternatively, the beam could rotate in azimuth while elevating slowly to trace out a helical-scan pattern. The main disadvantage of a radar having a pencil beam is that it usually take a relatively more time to cover the desired volume. The search time depends on the number of bits to be obtained from each target. More the number of hits per scan, the more accurate will be the angle measurement and the better the MTI performance. The time t_s required to scan an antenna of azimuth beamwidth θ_B and elevation beamwidth ϕ_B over a total azimuth angle θ_t^o and total elevation angle θ_t^o when n pulses are to be received from each resolution all (with a PRF f_r) is

$$t_s = \frac{n\phi_t \theta_t}{fr\,\phi_B \theta_B}$$

Elevation information may be obtained by **stacking** a number of narrow pencil beams in elevation and recording which beam contains the echo. Each of the stacked beams feeds an independent receiver and a separate transmitter might be used for each beams. By interpolating the voltages between adjacent beams of the stacked beam configuration, it is possible to obtain a more precise measurement of elevation angle.

Another technique to find elevation angle using from beam shape beam. Even though the broad beamwidth of a fan beam in elevation does not permit the measurement of the elevation angle to any degree of precision, it is possible to obtain a rough approximation of target height.

A technique of height finding employed is the **'V' Beam Radar.** This consists of two fan shape beams, are vertical and the other is tilted at some angle to the vertical as shown in Fig. 6.1.

The angle between the vertical beam and the slant beam is 45° as shown in Fig. 6.1. The time between observation of the same target in the two beams depends upon the target range and height of the target. The height of the target can be find by following relation,

$$h = \frac{R \sin \Delta\omega}{(1+\sin^2 \Delta\omega)^{1/2}} \quad \text{...(6.1)}$$

where h = Height of the target

R = Range of the target

$\Delta\omega$ = Azimuth rotation between beam = $\omega_s t_n$

ω_s = Azimuth rotation rate RPS

t_n = Time between observation in sec.

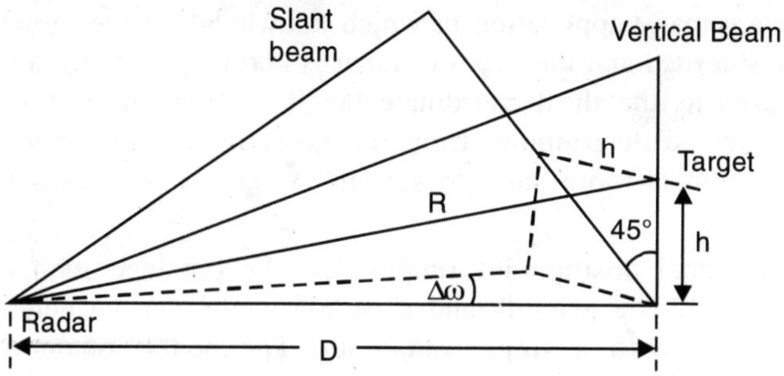

Fig. 6.1. **Geometry of the V-beam Radar.**

6.4. THEORETICAL ACCURACY OF RADAR MEASUREMENTS

The ability of a radar to detect the presence of an echo signal is fundamentally limited by the noise. Noise is the main limitations to accurate radar measurements. The parameters of radar applications are the range, or time delay, the range gets or Doppler velocity and the angle of arrival. The theoretical aspects of the extraction of information from radar signals are greatly benefitted by the theory of statistical parameter estimation. In this section, the theoretical accuracies of various radar measurements will be discussed. To simplifies the analysis, it is assumed that the signal is large compared with the noise. It is further assumed that signal to noise ratios is large enough to have the reliable detection decision. Furthermore, large signal to noise ratio are necessary for accurate measurements. It is also assumed that the error associated with a measurement of a particular parameter is independent of the errors in angle of the other parameters. The measure of error is the root mean square (rms) of the difference between the estimated value and the actual value.

The expression for the theoretical rms error ΔM of a radar measurement M have the following form :

$$\Delta M = \frac{KM}{\sqrt{\frac{2E}{N_0}}} \quad ...(6.2)$$

where K = Constant (unity approximately)

E = Received signal energy

N_0 = Noise power per unit bandwidth

Theoretical radar accuracies may be achieved by a variety of methods, which are as follows :

(*i*) Simple geometrical relationship between signal, noise and the parameter to be measured.

(*ii*) Inverse probability.

(*iii*) A suitably selected getting function preceded by a natural filter.

(*iv*) The estimate of the variance using the likelihood function.

6.5. AMPLITUDE, PHASE AND FREQUENCY OF A SINE WAVE

Let us consider a continuous sine wave $A \sin(2\pi ft + \phi)$,

where A = Amplitude

f = Frequency

ϕ = Phase

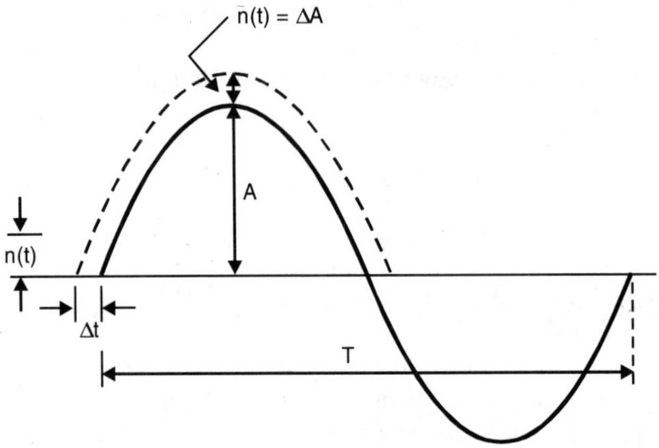

Fig. 6.2. One Period of Sine Wave—Dashed Curves Represents Signal Plus Noise.

One period or one cycle of the sine wave is shown in Fig. 6.2. The unwanted noise $[n(t)]$ causes the apparent amplitude shift from the true value of amplitude by an amount $\Delta A = n(t)$. The rms error in measuring amplitude is therefore

$$\delta A = (n^2)^{1/2} \qquad ...(6.3)$$

The relative error is

$$\frac{\delta A}{A} = \frac{1}{\left(\dfrac{A^2}{n^2}\right)^{1/2}} = \frac{1}{\left(\dfrac{2S}{N}\right)^{1/2}} \qquad ...(6.4)$$

where $\frac{S}{N}$ is the signal to noise power ratio.

The measurement of phase may be considered as the measurement of the time at which the waveform crosses the zero axis.

The error in estimating the time of a particular crossing is

$$\Delta t = \frac{x(t)}{\text{Slope of sine wave at zero crossing}} \qquad ...(6.5)$$

The slope of the sine wave is $2\pi fA$ at the point of zero crossing.

Hence

$$\delta t = \left[\overline{(\Delta t)^2}\right]^{1/2} = \frac{(\overline{n^2})^{1/2}}{2\pi fA} = \frac{1}{2\pi f\left(\dfrac{2S}{N}\right)^{1/2}} \qquad ...(6.6)$$

Since $\phi = 2\pi ft$, the phase error $f\phi = 2\pi f\delta t$ is

$$\delta\phi = \frac{1}{\left(\dfrac{2S}{N}\right)^{1/2}} \qquad ...(6.7)$$

The period 'T' is the time between two successive zero crossing of the same slope. The rms error in measuring the period will therefore be $\sqrt{2}$ times the single zero-crossing rms error, assuming that the two zero-crossing measurements are independent.

$$\delta T = \frac{\sqrt{2}}{2\pi f\left(\frac{2S}{N}\right)^{1/2}} = \frac{T}{2\pi\left(\frac{S}{N}\right)^{1/2}} \quad ...(6.8)$$

The error in period ΔT is related to the error in frequency (Δf) by $\Delta T = \frac{\Delta f}{f^2}$;

$$\delta f = \frac{f}{2\pi\left(\frac{S}{N}\right)^{1/2}} \quad ...(6.9)$$

The relative rms error in frequency and period are of the form

$$\frac{\delta f}{f} = \frac{1}{2\pi\left(\frac{S}{N}\right)^{1/2}} \quad ...(6.10)$$

$$\frac{\delta T}{T} = \frac{1}{2\pi\left(\frac{S}{N}\right)^{1/2}} \quad ...(6.11)$$

It is seen that all the errors such as phase, relative amplitude, period and frequency errors are depends upon the $\frac{S}{N}$ ratio.

6.6. RANGE ACCURACY (LEADING EDGE MEASUREMENT)

Actually the measurement of range is nothing but the measurement of time delay $\left(T_R = \frac{2R}{c}\right)$.

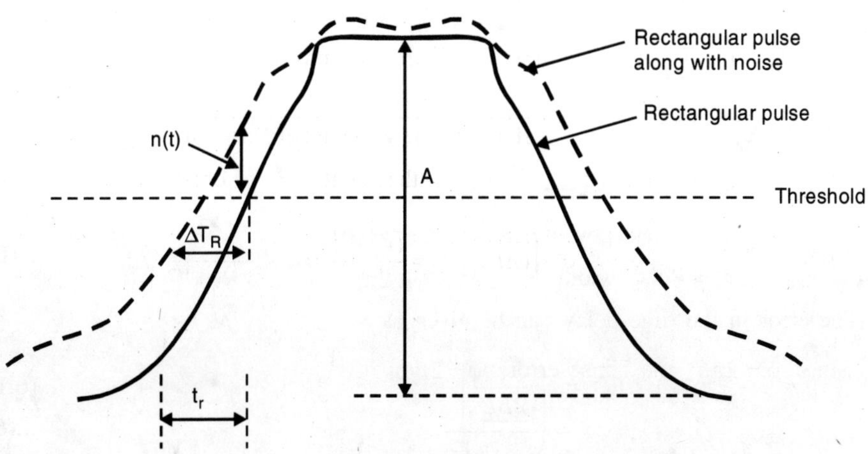

Fig. 6.3. Measurement of Time Delay using the Leading Edge of the Pulse.

where c = Velocity of light. One technique of determining range with a pulsed waveform is to measure the time at which the leading edge of the pulse crosses some threshold as shown in Fig. 6.3. The shape of the pulse is not perfectly rectangular (the size and decay times are not zero). The pulse shown by solid curve is the uncorrupted by noise. The noise effect the shape of the pulse and to shift the time of threshold crossing as shown by dashed curve.

The maximum slope of the leading edge of a pulse of amplitude 'A' at the point of a video filter is $\frac{A}{t_r}$, where t_r = sine time. For more $\frac{S}{N}$ ratio, the slope of the pulse affected by noise is essentially the same as the slope of the uncorrupted pulse. The slope of the pulse in noise = $\frac{n(t)}{\Delta T_R}$,

where $n(t)$ = Noise voltage,

ΔT_R = Error in time delay measurement

$$\Delta T_R = \frac{n(t)}{\frac{A}{t_r}} \qquad ...(6.12)$$

or

$$\left[\overline{(\Delta T_R)^2}\right]^{1/2} = \delta T_R = \frac{t_r}{\left(\frac{A^2}{n^2}\right)^{1/2}} = \frac{t_r}{\left(\frac{2S}{N}\right)^{1/2}} \qquad ...(6.13)$$

where $\frac{A^2}{n^2}$ = the video signal to noise ratio.

By Eqn (6.13), it shows that the video signal to noise power ratio is equal to twice the IF signal to noise power ratio, assuming a linear detector law and a large $\frac{S}{N}$ ratio.

If the bandwidth ('B') of the IF amplifier limits the rise time of the video pulse then $t_2 \approx \frac{1}{B}$.

Let $S = \frac{E}{\tau}$ and $N = N_0 B$

where E = Signal energy

N_0 = Noise power per unit bandwidth

τ = Pulse width

The error in the time delay can be given as

$$\delta T_R = \left(\frac{\tau}{\frac{2BE}{N_0}}\right)^{1/2} \qquad ...(6.14)$$

And if a similar independent time delay measurements is made at the trailing edge of the rectangular pulse, the two combined measurement will be improved by a factor of $\sqrt{2}$ as

$$\delta T_R = \left(\frac{\tau}{\frac{4BE}{N_0}} \right)^{1/2} \text{ rectangular pulse} \quad ...(6.15)$$

Equation (6.15) gives the rms delay error which is proportional to the rise time and is independent by the pulse width. An improvement in accuracy is obtained, therefore, by decreasing the rise time (increasing the BW) or increasing the S/N ratio. The selection of threshold level is important, for consistency in accurate measurement when only the leading edge is used.

6.7. TIME-DELAY ACCURACY AND EFFECTIVE BW

There are various methods based on likelihood ratio, inverse probability and other statistical analysis. By using these methods we will lead to the following expressions for the rms error in the measurement of the time delay :

$$\delta T_R = \frac{1}{\beta \left(\frac{2E}{N_0} \right)^{1/2}} \quad ...(6.16)$$

where β is called the effective bandwidth and which is defined as

$$\beta^2 = \frac{\int_{-\infty}^{+\infty} (2\pi f)^2 \, | S(f) |^2 \, df}{\int_{-\infty}^{+\infty} | S(f) |^2 \, df}$$

$$= \frac{1}{E} \int_{-\infty}^{+\infty} (2\pi f)^2 \, | S(f) |^2 \, df \quad ...(6.17)$$

β may be called an rms bandwidth.

***(i)* For rectangular Pulse :** When the pulse is perfect rectangular pulse *i.e.* zero rise time and zero fall time, for the effective bandwidth to $\beta = \infty$ and hence $\delta T_R = 0$ and range error is zero.

***(ii)* For Quasi-Rectangular Pulse :** The Fig. 6.4 shows that solid curve is the pulse shape that emerges out from the video low-pass filter of $BW\left(\frac{B_s}{2}\right)$ which corresponds to an IF filter with $BW(B_s)$. The Fig. 6.4 does not show a perfectly rectangular pulse, it is called the **quasi-rectangular pulse.** The rms time delay error in quasi-rectangular pulse is given by —

$$\delta T_R = \left(\frac{\tau_r}{2.1 \left(\frac{2E}{N_0} \right)^{1/2}} \right) \quad (BT_r \approx 1) \quad ...(6.18)$$

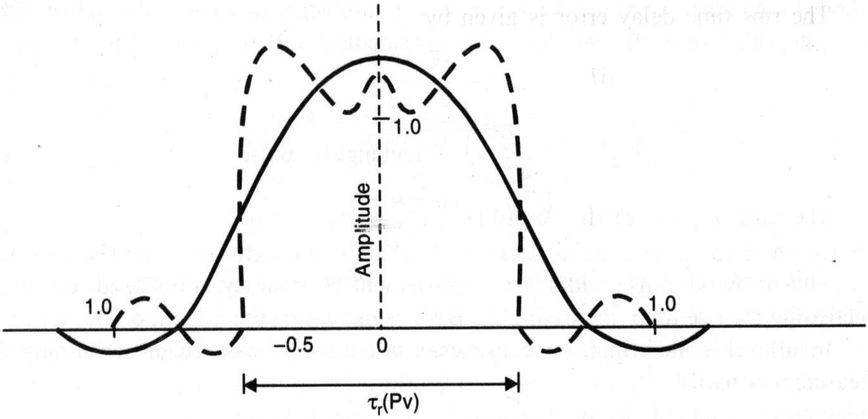

Fig. 6.4. A Quasi-Rectangular Pulse.

***(iii)* For Trapezoidal Pulse :** Sometimes a rectangular is approximated by a trapezoidal pulse with finite rise and fall times. If the width of the top of the trapezoid is τ_P and rise and fall time are t_r :

The rms error in time delay measurement is given as :

$$\delta T_R = \left(\frac{3\tau_P t_r + 2t_r^2}{\frac{12E}{N_0}} \right)^{1/2} \quad ...(6.19)$$

It is to be noted that the larger the width τ_P, the poorer is the accuracy.

***(iv)* For Triangular Pulse :** In case of triangular pulse the $\delta_P = 0$ and its base $= 2t_r = \tau_B$, then the rms error becomes

$$\delta T_R = \frac{\tau_s}{12^{1/2}\left(\frac{2E}{N_0}\right)^{1/2}} \quad ...(6.20)$$

***(v)* For Gaussian Pulse :** When the pulse is described by the function of Gaussian,

$$S(t) = \exp\left(-\frac{1.38t^2}{\tau^2} \right) \quad ...(6.21)$$

where δ = Half power pulse width.

Its rms time delay error is given as

$$\delta T_R = \frac{\tau}{1.18\left(\frac{2E}{N_0}\right)^{1/2}} = \frac{1.18}{\pi B\left(\frac{2E}{N_0}\right)^{1/2}} \quad ...(6.22)$$

where B is the half-power bandwidth of the Gaussian pulse spectrum.

***(vi)* Pulse with Uniform Spectrum :** The effective bandwidth of a waveform having uniform spectrum of width (B_u) is $\beta = \frac{\pi B_u}{\sqrt{3}}$.

The rms time delay error is given by

$$\delta T_R = \frac{\sqrt{3}}{\pi B_u \left(\frac{2E}{N_0}\right)^{1/2}} \quad ...(6.23)$$

The time wave is of the form of $\frac{\sin \pi B_u t}{\pi B_u t}$ waveform.

This is applied to a linear FM pulse compression waveform with large time bandwidth product.

In all the cases which we discussed till now, it was assumed that the radar measurement is based on the video waveform and not the RF or IF waveform. Range measurement using the RF or IF waveform are more accurate but they lead to ambiguous results. The ambiguities are in terms of half RF wavelength. The range information contained in the RF or IF signal is called **the fine-grain range information**. Figure 6.6 shows the appearance of the fine structure and it is a plot of the output signal amplitude as a function of time.

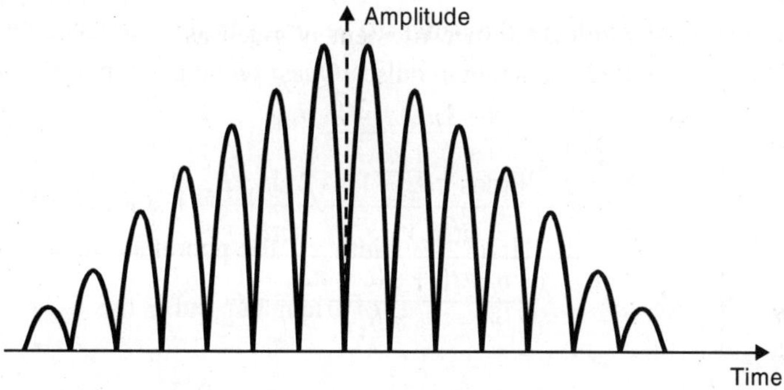

Fig. 6.5. Structure of the Time Grain Range Information.

6.8. ACCURACY OF FREQUENCY (OR DOPPLER VELOCITY) MEASUREMENT

The measurement of frequency in radar is related to the Doppler Frequency shift. We know that the radar component v_r of velocity can be found from the Doppler frequency shift $f_d = \frac{2v_r}{\lambda}$. The rms error in radial velocity is given by

$$\delta v_r = \left(\frac{\lambda}{2}\right)\delta f_d \quad ...(6.24)$$

where δf_d = rms error in the Doppler frequency.

By using the method of inverse probability, Manasse showed that the minimum rms error in the measurement of frequency is

$$\delta f = \frac{1}{\alpha\left(\frac{2E}{N_0}\right)^{1/2}} \quad ...(6.25)$$

where $$\alpha^2 = \frac{\int_{-\infty}^{\infty} (2\pi t)^2 s^2(t)\, dt}{\int_{-\infty}^{\infty} s^2(t)\, dt} \qquad ...(6.26)$$

$S(t)$ = Input signal as function of time.

Note that similarity between the definition of α and β, as well as between the expression of δf and δT_R. The parameter α is called the **effective time** duration of the signal. In radar, the measurement error specified by Eqn. (6.25) is that of the doppler frequency shift.

For a perfectly rectangular pulse the value of α^2 is $\frac{\pi^2\tau^2}{3}$; then the rms error in frequency is

$$\delta f = \frac{\sqrt{3}}{\pi\tau\left(\frac{2E}{N_0}\right)^{1/2}} \quad \text{rectangular pulse} \qquad ...(6.27)$$

The longer the pulse width (τ), the better the accuracy of the frequency measurement.

For a bandwidth limited rectangular pulse (Quasi-rectangular pulse), the value of α^2 is

$$\alpha^2 = \frac{\pi^2\tau^2}{3}\left[\frac{\frac{\cos \pi B_\tau - 3}{\pi B_\tau} - \frac{8(\cos \pi B_\tau - 1)}{(\pi B_\tau)^3} - \frac{2 \sin \pi B_\tau}{(\pi B_\tau)^2} + \sin(\pi B_\tau)}{\sin(\pi B_\tau) + \frac{(\cos \pi B_\tau - 1)}{\pi B_\tau}}\right] \qquad ...(6.28)$$

In the limit as $B_\tau \to \infty$, the value of α^2 approaches $\frac{\pi^2\tau^2}{3}$, which is the same as that of for the rectangular pulse.

For Trapezoidal Pulse : The rms error in the measurement of doppler frequency with a trapezoidal pulse is

$$\delta f = \frac{(3\tau_P + 2t_r)^{1/2}}{2\pi\left(\frac{\tau_P^3}{4} + \frac{\tau_P^2 t_r}{2} + \frac{\tau_P t_r^2}{2} \frac{t_2^3}{5}\right)^{1/2}\left(\frac{2E}{N_0}\right)^{1/2}} \qquad ...(6.29)$$

where, t_r = rise and full time

τ_P = Width across the top of the trapezoid

For triangular pulse : For the triangular pulse, we get $\tau_P = 0$ and $2t_r = \tau_B$. The rms error is

$$\delta f = \frac{(10)^{1/2}}{\pi T_B\left(\frac{2E}{N_0}\right)^{1/2}} \qquad ...(6.30)$$

The rms error in frequency for a Gaussian-shaped pulse is

$$\delta f = \frac{1.18}{\pi\tau\left(\frac{2E}{N_0}\right)^{1/2}} = \frac{B}{1.18\left(\frac{2E}{N_0}\right)^{1/2}} \qquad ...(6.31)$$

The time delay and frequency error expressions obtained in this section apply for a single observation. And when more than one independent observation is made, the resultant error may be found by continuing the errors in the usual manner.

Uncertainty Relation : The so called 'uncertainty' relation of radar describes that the product of the effective bandwidth β occupied by a signal waveform and the effective time duration α must be greater than or equal to π, which is given as

$$\beta\alpha \geq \pi \qquad ...(6.32)$$

Equation (6.32), sometimes been referred to as the **radar uncertainty principle**. This relationship may be derived from the definition of β and α given by earlier equation and by applying the Schwartz in equality.

The use of word 'uncertainity' is a monomer, for there is nothing considerations. Uncertain about the uncertainty relation Eqn. (6.32). It states that a narrow waveform yields a wide spectrum and a wide waveform fields a narrow spectrum and both the time waveforms and the frequency spectrum cannot be made arbitrarily small simultaneously.

By using Eqn. (6.32), the accuracy with time delay and frequency may be measured simultaneously. The product of rms time delay error and the rms frequency error is

$$\delta T_R \delta f = \frac{1}{\beta\alpha\left(\frac{2E}{N_0}\right)} \qquad ...(6.33)$$

Now substituting this in equality of eqn. (6.33)

$$\delta T_R \delta f \leq \frac{1}{\pi\left(\frac{2E}{N_0}\right)} \qquad ...(6.34)$$

The above Eqn. (6.34) states that the time delay and the frequency may be simultaneously measured to as small a theoretical error as one desires by designing the radar to yield a sufficiently large ratio of signal energy (E) to noise power per cycle (N_0) or for constant $\frac{E}{N_0}$, to select a radar waveform which result in a large value of βα. Large βα products requires waveforms long in duration and of wide bandwidth.

Angular Accuracy : The measurement of angular position is the measurement of the angle of arrival of the equiphase wavefront of the echo signal. Theoretically, the rms error of the angle measurement may be desired in the same manner similar to the derivations of time (range) and frequency errors. The similarity between the angular error and the time delay or frequency error comes about because the fourier transform describes the relationship between the radiation pattern and the aperture distribution of an antenna in a manner similar to the relationship between the time waveform and its frequency spectrum.

First, we will discuss the angular error in one co-ordinate plane for the simplicity. It is assumed that S/N ratio is large and that the noise can be described by the Gaussian probability-density function.

Let us consider a linear in phase receiving antenna of length *D* or a rectangular receiving antenna having aperture of width *DA* as shown in Fig. 6.7. The amplitude distribution across the aperture as a function of *x* is denoted *A*(*x*). The voltage as a function of the angle θ (one dimension radiation pattern) in the *xz* plane is proportional to

$$G_v(\theta) = \int_{-D/2}^{+D/2} A(x) \exp\left(j2\pi \frac{x}{\lambda} \sin\theta\right) dx \qquad ...(6.35)$$

When sin θ = θ (the angle θ is small), is recognized as an Inverse Fourer transform

$$G_v(\theta) = \int_{-D/2}^{D/2} A(x) \exp \frac{j2\pi x\theta}{\lambda} dx \qquad ...(6.36)$$

This is resembles to the inverse Fourier transform relating the frequency spectrum *S*(*f*) and time waveform *S*(*t*),

$$S(t) = \int_{-\infty}^{\infty} S(f) \exp\left(j\,2\pi f_t\right) df \qquad ...(6.37)$$

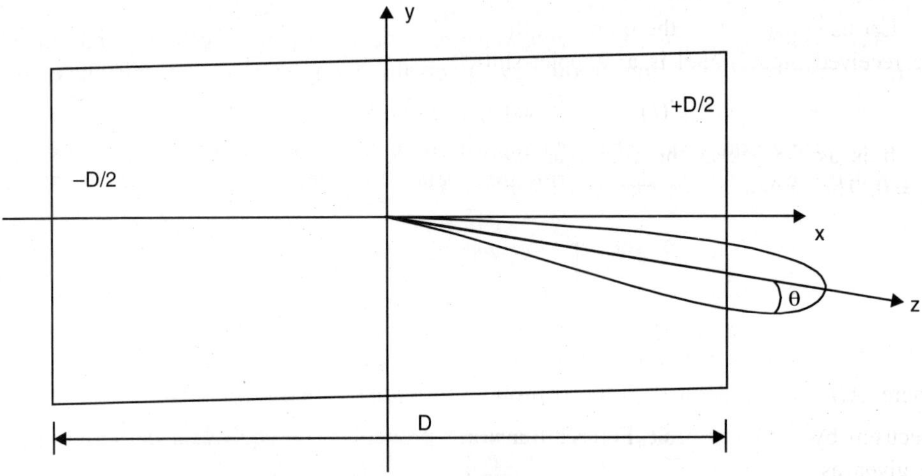

Fig. 6.6. Rectangular Receiving Aperture by Width '*D*'.

The theoretically rms error can be determined by analogy to the time-delay accuracy as,

$$f\left(\frac{\theta}{\lambda}\right) = \frac{1}{\gamma\left(\frac{2E}{N_0}\right)^{1/2}} \qquad ...(6.38)$$

where γ = The effective aperture width and can be defined as

$$\gamma^2 = \frac{\int_{-\infty}^{\infty} (2\pi x)^2 \,|A(x)|^2\, dx}{\int_{-\infty}^{\infty} |A(x)|^2\, dx} \qquad ...(6.39)$$

Now the theoretically angular error of an antenna with a rectangular amplitude distribution across the aperture is

$$\delta\theta = \frac{\sqrt{3}\lambda}{\pi D\left(\frac{2E}{N_0}\right)^{1/2}} \quad ...(6.40)$$

or
$$\delta\theta = \frac{0.628\theta_B}{\left(\frac{2E}{N_0}\right)^{1/2}} \quad ...(6.41)$$

where θ_B = Half power beamwidth (radian).

Here the relative error $\left(\frac{\delta\theta}{\theta_B}\right)$ is a function of $\frac{E}{N_0}$ only.

6.9. AMBIGUITY DIAGRAM

The ambiguity function (AF) is an analytical important tool for designing of the waveform and analysis the response of a waveform paired with its matched filter. The AF is very useful to observe resolution, side lobe behaviour and ambiguities in Doppler and range for a particular waveform.

Let us consider that the output of the matched filter for a given waveform $x(t)$ and the received input signal is a Doppler-shifted response

$$x'(t) = x(t)\exp(j2\pi F_D t)$$

It is assumed that the filter has unity gain ($\alpha = 1$) and constructed to peak at $T_m = 0$. The output of the matched filter will be

$$y(t;\, F_D) = \int_{-\infty}^{\infty} x(s)\exp(j2\pi F_D s)\, x^*(s-t)\, ds$$

$$= \hat{A}(t, F_D) \quad ...(6.26)$$

where $\hat{A}(t, F_D)$ is defined as the **complex ambiguity function.** In term of the signal spectrum by applying basic Fourier transform properties, an equivalent definition can be given as

$$\hat{A}(t, F_D) = \int_{-\infty}^{\infty} X^*(F)\, X(F - F_D)\exp(j2\pi F t)\, dF \quad ...(6.27)$$

The ambiguity function is described as the magnitude of $\hat{A}(t, F_D)$ *i.e.*,

$$A(t,\, F_D) = \left|\hat{A}(t, F_D)\right| \quad ...(6.28)$$

The ambiguity function has two variables, the time delay relative to the expected matched filter peak output, and the mismatch between the Doppler shift for that the filter was constructed, and that which is actually received.

There are three properties of the ambiguity function which are of main interest. They are as follows :

(*i*) If the waveform has energy E, then

$$\left|A(t, F_D)\right| \le \left|A(0,0) = E\right| \quad ...(6.29)$$

It states that when the filter is matched in both Doppler and range the response will be maximum and if the filter is not matched the response is always less than the maximum.

(*ii*) The second property of ambiguity function is that the total area under any AF is constant and it is given as

$$\int_{-\infty}^{\infty}\int_{-\infty}^{\infty} |A(t, F_D)|^2 \, dt \, dF_D = E^2 \qquad \text{...(6.30)}$$

According to second property of AF, that the energy from the waveforms cannot be removed from one portion of the ambiguity surface until placing it somewhere else. The property is derived from law of conservation of energy.

(*iii*) The third property is a symmetry relation and given as

$$A(t, F_D) = A(-t, -F_D) \qquad \text{...(6.31)}$$

After analysing the properties of AF, now we should discuss an ideal ambiguity function. An ideal ambiguity function varies according to the intent of the system design. The ideal AF would equal unity near (0, 0) and zero elsewhere and such an AF is referred to as a **thumb track** AF. Figure 6.7 shows the thumb track ambiguity function, which shows a single central peak and the remaining energy spread uniformly throughout the delay Doppler plane. In Figure 6.7, there is only one narrow central peak that means good resolution in both range and doppler and there no other secondary peak implies that there is no ambiguities in range and doppler. These features are useful for a system designed to make high resolution measurements of targets in range and doppler. An ideal AF depends on the use to which waveform to designed.

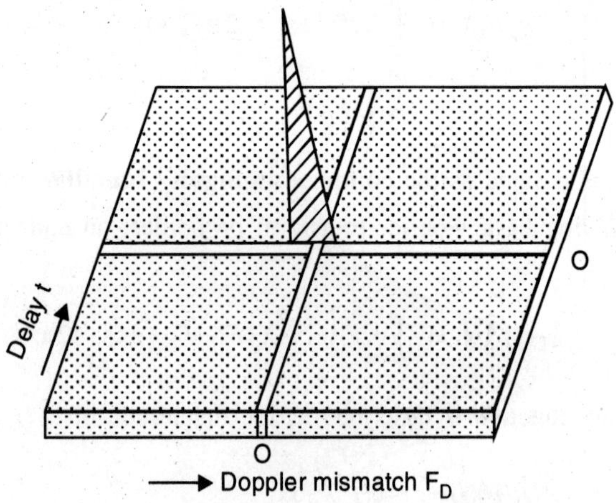

Fig. 6.7. Thumb track AF.

6.9.1. Ambiguity Function of Simple Matched Pulse

To discuss the ambiguity function of simple pulse, let us consider a simple pulse, centered on the origin and normalized so that to have unit energy *i.e.* $E = 1$,

$$x(t) = \frac{1}{\sqrt{\tau}} \qquad \left(\frac{-\tau}{2} \le t \le \frac{\tau}{2}\right) \qquad \text{...(6.32)}$$

By using Eqn. (6.26) of previous section, for $t > 0$

$$\hat{A}(t, F_D) = \int_{-\tau/2+t}^{\tau/2} \frac{1}{\tau} e^{j2\pi F_D s}\, ds$$

$$= \frac{e^{j2\pi F_D \frac{\tau}{2}} - e^{j2\pi F_D\left(\frac{-\tau}{2}+t\right)}}{\tau\, j2\pi F_D}$$

$$= \frac{1}{\tau\, j2\pi F_D} \cdot e^{j2\pi F_D \frac{\tau}{2}} \left\{ \exp\left[j2\pi F_D\left(\frac{\tau}{2} - \frac{t}{2}\right)\right] - \exp\left[-j2\pi F_D\left(\frac{\tau}{2} - \frac{t}{2}\right)\right]\right\} \quad \text{...(6.33)}$$

For $t > 0$, the magnitude of AF is given in Eqn. (6.33)

$$A(t, F_D) = \left|\hat{A}(t, F_D)\right| = \left|\frac{\sin(\pi F_D(\tau - t))}{\tau\pi F_D}\right| \quad 0 \le t \le \tau \quad \text{...(6.34)}$$

Now by applying Eqn. (6.26) of previous section. For $t < 0$, we get similar result as in Eqn. (6.34) except with quantity $(\tau - t)$ is replaced by $(\tau + t)$.

The complete ambiguity function of the simple pulse is given as

$$A(t, F_D) = \left|\frac{\sin[\pi F_D(\tau - |t|)]}{\tau\pi F_D}\right| \quad -\tau \le t \le \tau \quad \text{...(6.35)}$$

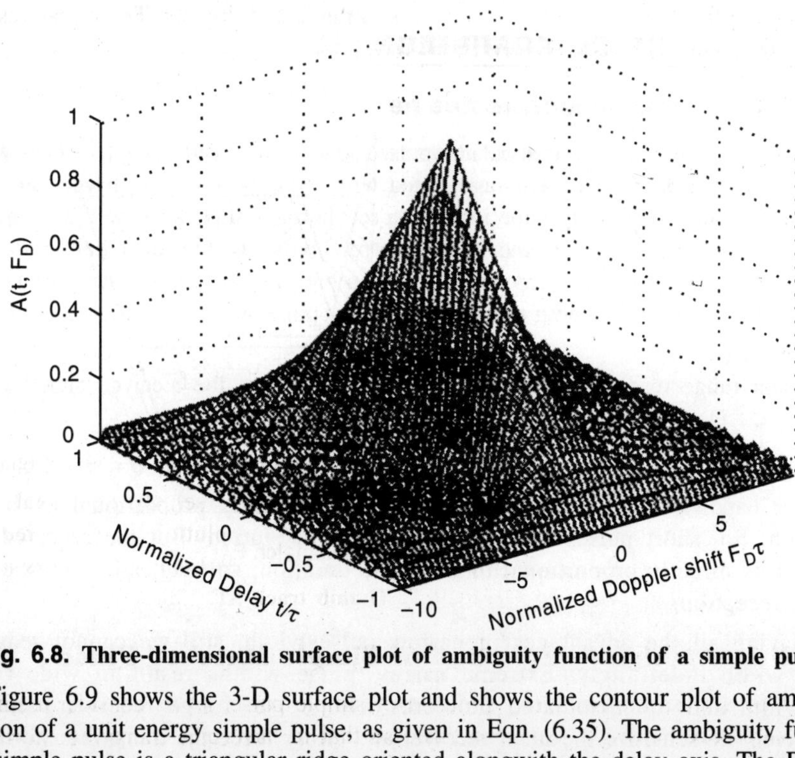

Fig. 6.8. Three-dimensional surface plot of ambiguity function of a simple pulse.

Figure 6.9 shows the 3-D surface plot and shows the contour plot of ambiguity function of a unit energy simple pulse, as given in Eqn. (6.35). The ambiguity function of a simple pulse is a triangular ridge oriented alongwith the delay axis. The Doppler mismatch order of $\frac{1}{\tau}$ Hz or more severely reduce and spread the matched filter output peak.

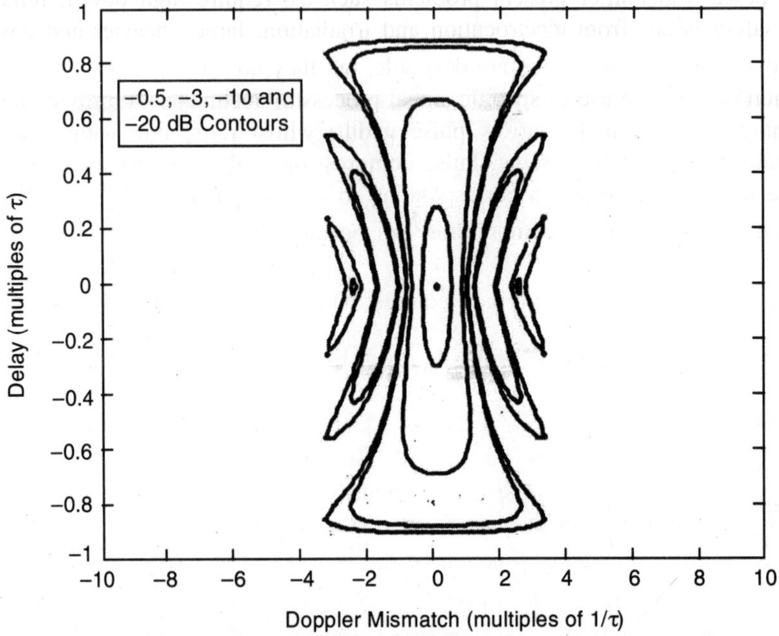

Fig. 6.9. Contour plot of the simple ambiguity function of simple pulse.

6.10. PULSE COMPRESSION

Why Pulse compression is needed:

When we have discussed about the pulse radar in the earlier chapter, there we have described numbers of advantages associated with radar operating with very short pulse widths. The advantages are superior range resolution, range accuracy, and minimum range. In addition to those advantages, narrow pulse widths also help radar when operating in a cluttered environment. Radar also has an ability to perform limited target classification if operating with sufficiently narrow pulse width or sufficiently fine range resolution.

Radar range resolution depends on the bandwidth of the received signal,

$$\rho = c\tau/2 = c/2B;$$

where, c = speed of light, ρ = range resolutions, τ = pulse width and B = signal bandwidth.

The bandwidth of a time-gated sinusoid is inversely proportional to the pulse duration. So, short pulses are better for range resolution. But the received signal strength is directly proportional to the pulse duration; so long pulses are better for signal reception.

Having all the advantages of narrow pulse width, still we cannot reduce the pulse width indefinitely. Extreme narrow pulse widths result in wide receiver bandwidth and the associated problems with noise. Large receiver bandwidths effectively de-sensitize the radar receiver and either force the transmitter to transmit higher levels of peak power to compensate, or accept the consequential reduction in range. And there are always limits on the amount of peak power available from the transmitter, and invariably a reduction in pulse width leads to a reduction in the maximum range of a radar.

High power transmitter present problems such as, require high power, reliability problem, safety issues from electrocution and irradiation, larger, heavier and costlier.

In short, narrow pulse widths are desirable, but they are not always feasible. Pulse compression radar make use of specific signal processing techniques to provide most of the advantages of extremely narrow pulse widths whilst remaining within the peak power limitations of the transmitter. Pulse compression is the compromise transmits a long pulse that has a bandwidth corresponding to a short pulse.

The advantages of pulse compression can be understood well with the following example;

Desired resolution, ρ = 15 cm, Required bandwidth = 1 GHz, Required pulse energy, E = 1 mJ.

$$E(J) = P(W)\cdot\tau(s)$$

As per Brute force approach

Raw pulse duration, $\tau = 1$ ns (10^{-9}s) Required transmitter power, P = **1 MW!**

As per Pulse compression approach

Pulse duration, $\tau = 0.1$ ms (10^{-4}s) Required transmitter power, P = **100 W**

So, now we can explain the pulse compression technique in broader ways.

Pulse Compression is the processing of a wideband, coded signal pulse, of initially long time duration and low range resolution, to result in an output pulse of time duration corresponding to the reciprocal of the bandwidth and, hence, higher range resolution, and with approximately the same pulse energy.

In principle, the process of pulse compression is the by effect when the signal with intra pulse modulation is processed in the matched filter to maximize signal to noise ratio. Since the matched filter does not preserve the initial shape of the waveform at its output, but on the contrary, distorts it to obtain the benefit of superposition of the maxima of different harmonics (to get the highest possible signal to noise ratio), the resultant output waveform compresses in time in comparison with the input waveform by value of the pulse compression ratio.

This useful property of matched filter processing gives the benefit of radiating long pulses on transmit and, hence obtaining efficient use of power capability, and simultaneously obtaining short pulses on receive and, hence obtaining good range resolution when employing pulse compression waveforms. The side effect of pulse compression is appearance of range or time side lobes that can mask nearby echoes, requiring the use of special measures for their suppression.

There are two basic ways to implement intra pulse modulation, coding either the frequency or the phase of the transmitted pulses, resulting in frequency coded or phase coded waveforms, which are the basic types of waveforms used in modern radars.

6.10.1. Concept of Operation

The block diagram of pulse compression radar is almost similar to the pulse radar, which is shown in Fig. 6.10. The block diagram shows the frequency modulator responsible for generating the frequency modulated (chirp) pulse. The pulse compression filter is an example of a matched-filter because the filter is specially designed to recognize the characteristics of the transmitted pulse as they are returned to the receiver in the form of reflected pulses. The filter has been matched to the transmitted waveform.

Received pulses with similar characteristics to the transmitted pulse are recognized by the matched filter where other received signals pass relatively unnoticed by the receiver.

The basic methods of implementation of pulse compression are analog and digital pulse compression techniques. The main advantages of pulse compression, leading to wide usage of this technique in modern radar are the following, increased detection capability inherent in long, high energy pulses is combined with increased resolving capability inherent in short pulses, generation of high peak power common in short pulse systems can be avoided and more efficient use of average power can be obtained without increasing pulse repetition frequency, which would result in decreasing the unambiguous range; since the long pulse is used on transmit, increased Doppler resolution is possible, when coded waveforms are used, a radar is less vulnerable to interference. The cost to be paid for these advantages is greater complexity relative to simple pulse transmissions, complexity of pulse compression waveform generation and processing, all increasing the cost of the radar system.

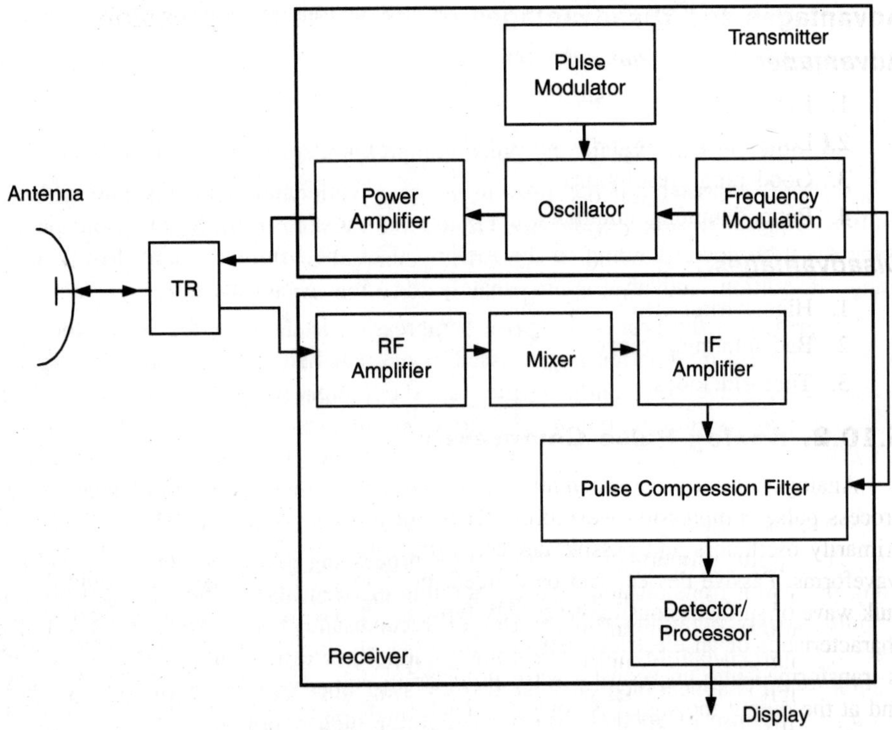

Fig. 6.10. The Block Diagram of Pulse Compression Radar.

The ability of the receiver to improve the range resolution over that of the conventional system is called the *pulse compression ratio (PCR)*. For example, a pulse compression ratio of 50:1 means that the system range resolution is reduced by 1/50 of the conventional system. The pulse compression ratio can be expressed as the ratio of the range resolution of an unmodulated pulse of length τ to that of the modulated pulse of the same length and bandwidth B.

$$PCR = \frac{\left(c_0 \cdot \frac{\tau}{2}\right)}{\left(\frac{c_0}{2B}\right)} = B.\tau \quad ...(6.27)$$

This term is described as *Time-Bandwidth-product* of the modulated pulse and is equal to the Pulse Compression Gain, PCG, as the gain in SNR relative to an unmodulated pulse. Alternatively, the factor of improvement is given the symbol PCR, which can be used as a number in the range resolution equation, which now achieves :

$$R_{res} = c_0 \cdot \left(\frac{\tau}{2}\right) . PCR = \frac{c_0}{2B} \quad ...(6.28)$$

The compression ratio is equal to the number of sub pulses in the waveform, *i.e.*, the number of elements in the code. The range resolution is therefore proportional to the time duration of one element of the code. The radar maximum range is increased by the fourth root of PCR.

The minimum range is not improved by the process. The full pulse width still applies to the transmission, which requires the duplexer to remain aligned to the transmitter throughout the pulse. Therefore $R_{\min}$ is unaffected.

Advantages and Disadvantages of the Pulse Compression

Advantages:

1. Higher maximum range
2. Lower pulse-power therefore suitable for Solid-State-amplifier
3. Good range resolution
4. Better jamming immunity.

Disadvantages:

1. High wiring effort
2. Bad minimum range
3. Time-sidelobes.

6.10.2. Analog Pulse Compression

Analog Pulse Compression involves the use of analog methods to generate and process pulse compression waveforms. The main techniques incorporate active devices, primarily oscillators and passive devices, primarily different delay lines for linear FM waveforms. Passive devices can be divided into two general classes, ultrasonic devices bulk wave or surface acoustic wave SAW types and electrical devices using the dispersive characteristics of an electrical network. In ultrasonic devices the input electrical signal is transformed into an acoustic wave propagating through the medium at sonic speed, and at the output the signal is converted back to an electrical waveform. Longer delays may be achieved than with purely electrical devices of comparable size, as the wave travels at sonic speed.

The most popular technique for linear FM waveforms is the SAW delay line. In SAW technology the energy is concentrated in a surface wave, making it much more efficient than bulk wave devices where the wave propagates through the crystal. The main limitation of bulk devices is the necessity to arrange the coupling between acoustic medium and electrical signal, typically with transducers inserting high losses currently interdigital transducers are considered to transform an electrical signal most efficiently to acoustic energy and vice versa. Electrical networks with dispersive characteristics are typically electrical delay networks having a linear delay versus frequency characteristics. The main characteristics of analog pulse compression devices for linear FM waveforms are given in the Table 6.1.

Table 6.1.

	B, MHz	*T*, μs	*BT*	f_0, MHz	Typical loss, dB	Typical spurious, dB
Aluminum strip delay line	1	500	200	5	15	–60
Steel strip delay line	20	350	500	45	70	–55
All-pass network	40	1000	300	25	25	–40
Perpendicular diffraction delay line	40	75	1000	100	30	–45
Surface-wave delay line	40	50	1000	100	70	–50
Wedge-type delay line	250	65	1000	500	50	–50
Folded-type meander line	1000	1.5	1000	2000	25	–40
Waveguide operated near cutoff	1000	3	1000	5000	60	–25
YIG crystal	1000	10	2000	2000	70	–20

The main characteristics of analog pulse compression devices for linear FM waveforms.

The main disadvantages of analog pulse compression are the instability of network parameters, resulting in poor repeatability of compressed waveforms, and larger size than digital circuits of comparable performance.

6.10.3. Digital Pulse Compression

Digital Pulse Compression involves use of digital technology for generating and processing pulse compression waveforms. It has some distinct advantages over analog pulse compression technique. It is more flexible, it allows implementation of different types of pulse compression waveforms binary and polyphase code waveforms, linear and nonlinear FM waveforms or even other exotic waveforms at designer's discretion. It can operate with signals of different durations, the same implementation can be used to handle multiple types of waveforms; and digital waveform generators are very stable devices.

A system for digital implementation of a binary coded waveform compression employing in phase and quadrature channels often called *homodyne or zero IF* is shown in Fig. 6.11. In this system the phase of each transmitted binary element is defined by the code generator and is equal to 0 or 180° with respect to the local oscillator signal, refer Fig. 6.12.

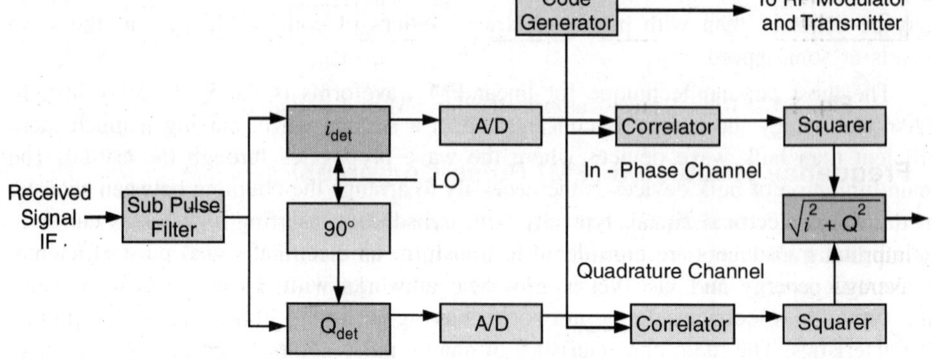

Fig. 6.11. A block-diagram of Digital pulse compression for binary coded waveform.

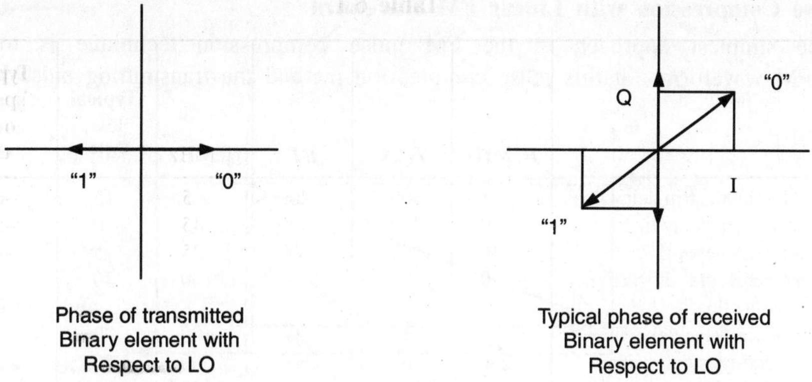

Fig. 6.12. Digital pulse compression for binary coded waveform.

The phase of the received signal is shifted in respect to the LO by an amount depending on the range and velocity of the target. Each correlator may consist of several correlators depending on the number of quantization bits in the digitized signal and only one channel instead of two channels can be used that will insert an average loss about 3 dB.

A typical scheme that may be used to generate digital frequency coded or polyphase coded waveforms is shown in the Fig. 6.13. The development of the digital pulse compression approach is closely tied with the general development of the digital technology and will develop with the further increase of memory unit's capability and computational speed. Nowadays the main limitation is the upper limit of frequency available for digital devices, so digital approach is restricted in bandwidth under about 100 MHz One of the ways to increase the bandwidth limitation for linear FM waveforms is to use frequency multiplication combined with stretch processing. The popular digital technique for pulse compression is the Fast Fourier Transform.

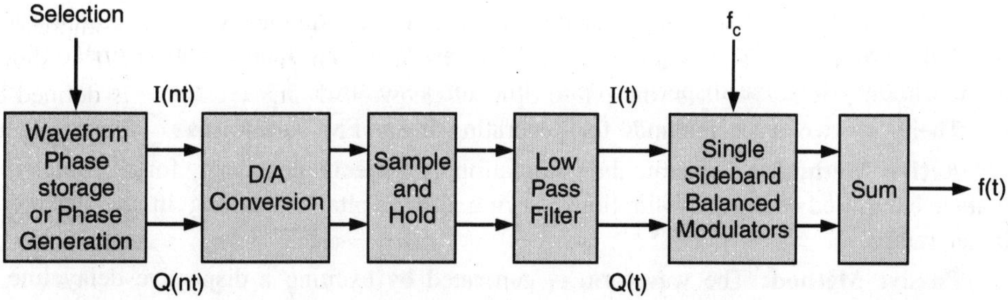

Fig. 6.13. Digital pulse compression for FM and polyphase coded waveforms.

Frequency Modulated [FM] Pulse Compression

Frequency Modulated [FM] Pulse Compression is one of the basic techniques for improving the range resolution of radar while maintaining adequate duty factor and average power. There are two type of FM pulse compression.

1. Pulse Compression with Linear FM Waveform.
2. Pulse Compression with Non Linear FM Waveform.

1. Pulse Compression with Linear FM Waveform

The simplest approach in the FM pulse compression technique is to use linear FM waveforms, at this pulse compression method the transmitting pulse has a

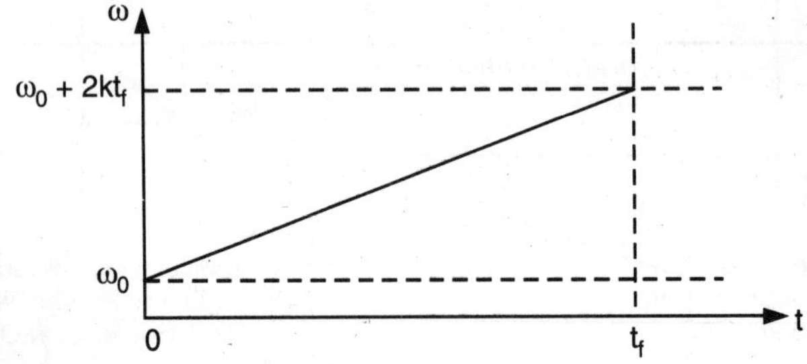

Fig. 6.14. Frequency vs. time for linear FM waveform.

linear FM waveform. The meaning of linear FM waveform is a frequency modulated waveform in which the carrier frequency varies linearly in time for some specified period for CW radar or within the pulsewidth for pulsed radar. To obtain this waveform, the phase must have a quadratic dependence on time. The waveform voltage can be written as :

$$v(t) = v_0 \cos \varphi(t) = v_0 \cos [\omega_0 t + \theta(t)] \quad \text{...(6.29)}$$

where, v_0 is the amplitude, ω_0 is carrier frequency, and $\theta(t)$ is the phase, when $\theta(t) = kt^2$, the instantaneous frequency; $\omega(t) = \dfrac{d\varphi(t)}{dt} = \omega_0 + 2kt$ varies linearly.

This is one of the oldest and most frequently used waveforms in both CW and pulse radar applications. In FM CW radars it makes possible the determination of target range as well as Doppler.

In pulse radar the frequency of this pulse compression waveform increases or decreases linearly over the duration of the pulse shown in Fig. 6.15, providing intrapulse modulation. A variant of the linear FM waveform is the linear step FM waveform, in which the frequency versus time function is not the continuous one, but comprises N equal steps as shown in Fig. 6.15. Such a waveform can approximate a linear FM pulse without using the dispersive delay line for generation.

There are two basic methods for generating linear FM waveforms :

Active Methods : Generate the modulation by a special generator, for example, a voltage controlled oscillator, with time varying drive voltage resulting in the desired linear ramp.

Passive Method: The waveform is generated by exciting a dispersive delay line with an impulse.

The main advantages of the linear FM waveform :

(*i*) It is quite insensitive to Doppler shifts.

(*ii*) It is the easiest to generate, a variety of hardware being available to form and process it.

The main disadvantages of linear FM waveforms are :

(*i*) It has range Doppler cross coupling, resulting in measurement errors unless one of the coordinates (range or Doppler) is determined.

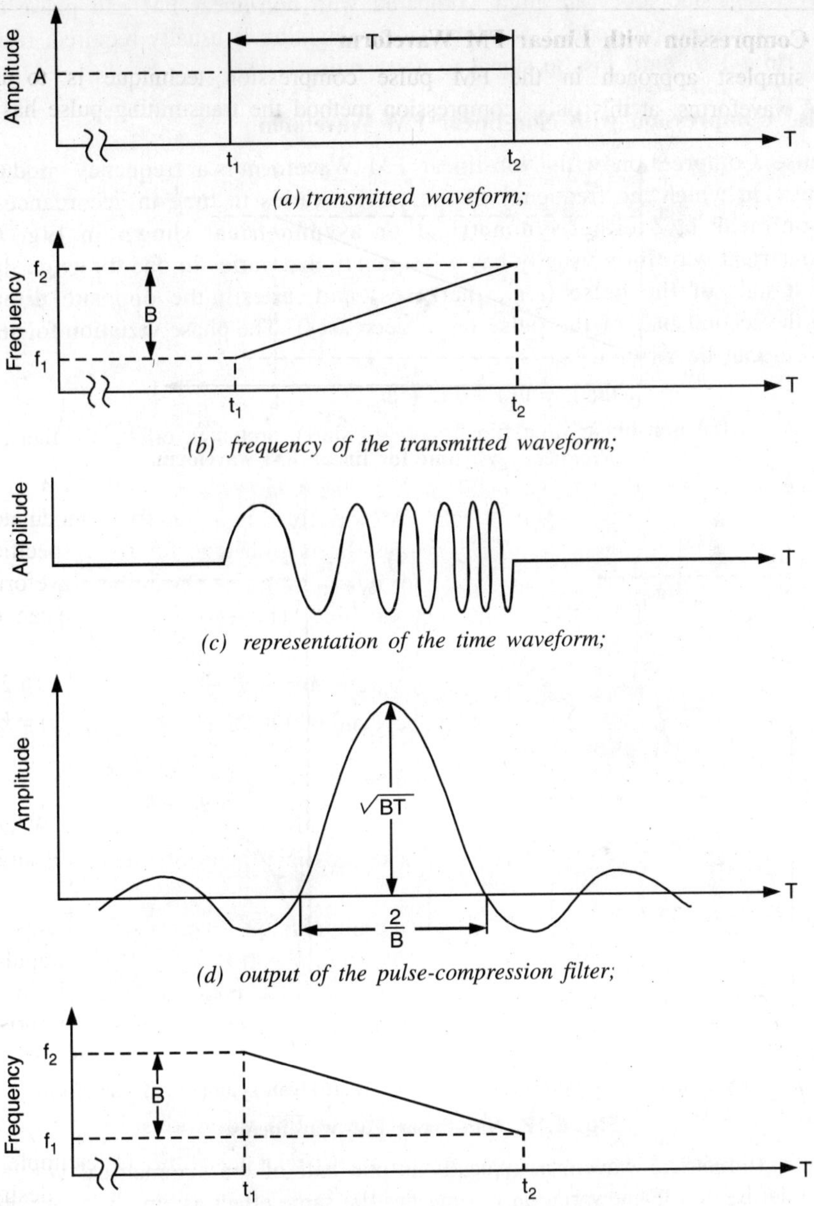

(a) *transmitted waveform;*

(b) *frequency of the transmitted waveform;*

(c) *representation of the time waveform;*

(d) *output of the pulse-compression filter;*

(e) *same as (b) but with decreasing frequency*

Fig. 6.15. Linear FM waveform.

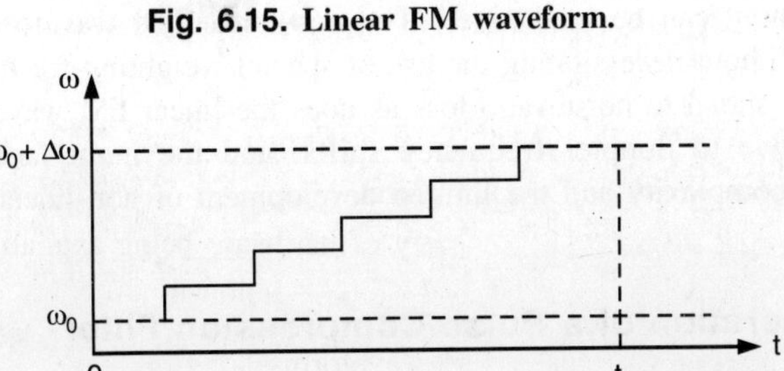

Fig. 6.16. Frequency vs. time for stepped FM waveform.

(*ii*) Range side lobes are high, compared with non-linear FM and phase coded waveforms. To reduce side lobe level, weighting is usually required, resulting in a 1 to 2 dB loss in signal to noise ratio.

2. Pulse Compression with Non-linear FM Waveform

Pulse Compression with Non-linear FM Waveform is a frequency modulated waveform in which the frequency of the carrier varies in time in accordance with any nonlinear law, either symmetrical or asymmetrical shown in Fig. 6.17. A symmetrical waveform usually has a frequency that varies in one direction during the first half of the pulse (*e.g.*, increases) and varies in the opposite direction during the second half of the pulse (*e.g.*, decreases). The phase variation for such a waveform can be written

$$\Phi(t) = \omega_0 t + kt^2 + \omega_n$$

where, $\omega_n t$ is the non-linear time function (*e.g.*, sin(t), arctan(t), erf(t), or other).

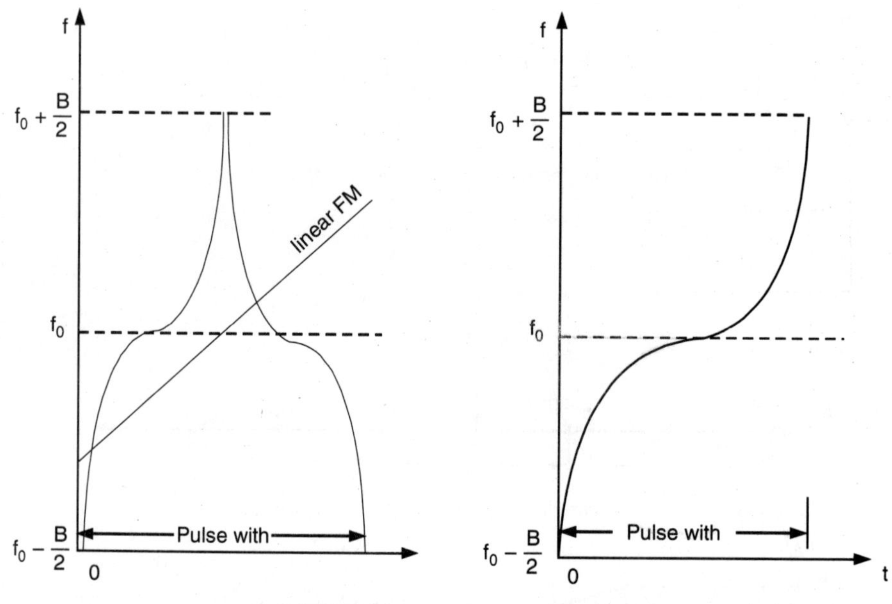

(*a*) Symmetrical waveform (*b*) Non-symmetrical waveform

Fig. 6.17. Non-linear FM waveforms.

The asymmetrical waveform typically is one half of the symmetrical waveform. In principle the non-linear variation in time has the same effect as amplitude weighting of the transmitted signal spectrum, but a rectangular pulse shape, compatible with efficient transmitter operation, can be maintained. The non-linear FM waveform has very low range sidelobes without necessitating the use of special weighting for their suppression, and hence has no signal to noise ratio loss as does the linear FM waveform. However, it is more sensitive to doppler frequency shifts, and the main factor affecting its acceptance is its complexity and the limited development of non-linear FM generation techniques.

Principles Operation of a Pulse Compression Filter

Now it is very important to discuss the principles operation of a pulse compression filter.

The compression filter is simply dispersive delay lines with a delay, which is a linear function of the frequency. The compression filter allows the end of the pulse to catch up to the beginning, and produces a narrower output pulse with higher amplitude. Filters for linear FM pulse compression radars are now based on two main types digital processing with (A/D) conversion, and Surface acoustic wave devices (SAW).

The output of the compression filter consists of the compressed pulse accompanied by responses at other times, called time or range sidelobes. The Fig. 6.18 shows a view of the compressed pulse of a chirp radar at an oscilloscope and at a PPI scope sector.

Amplitude weighting of the output signals may be used to reduce the time sidelobes to an acceptable level. Weighting on reception only results a filter mismatch and some loss of signal to noise ratio.

The sidelobe levels are an important parameter when specifying a pulse compression radar, the application of weighting functions can reduce time sidelobes to the order of 30 dB's.

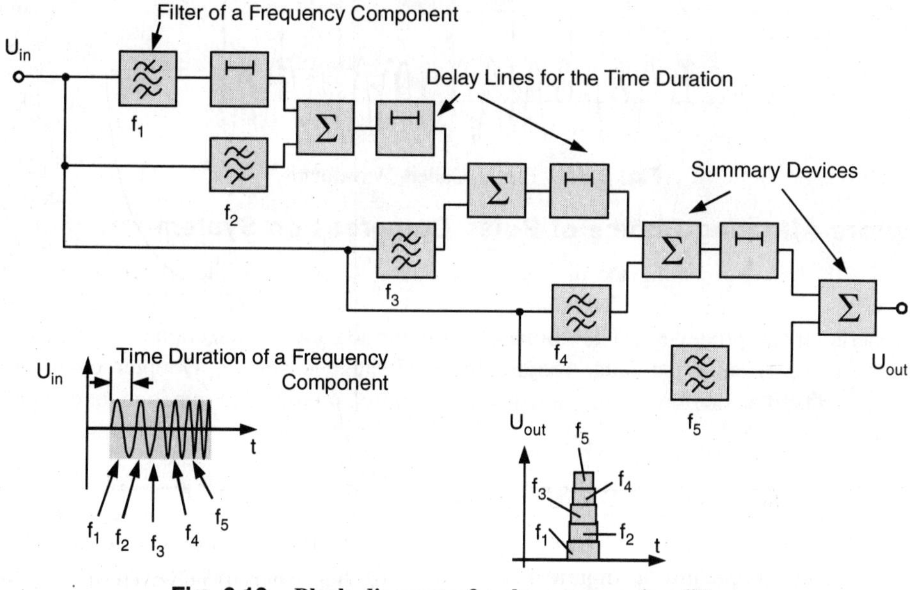

Fig. 6.18. Block diagram of pulse compression filter.

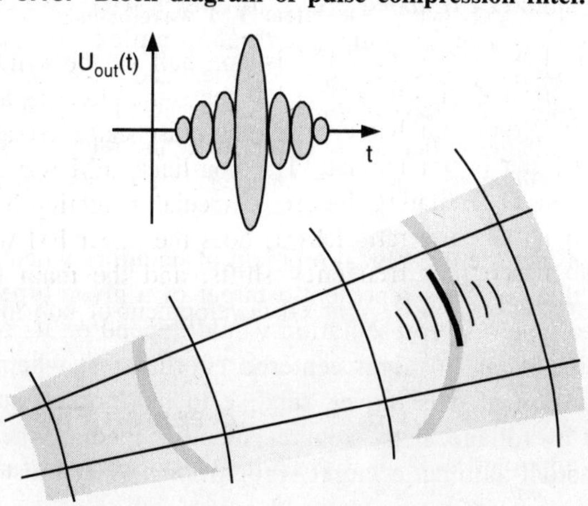

Fig. 6.19. The output the sidelobe.

Phase Coded Waveforms Pulse Compression

Phase coded waveforms differ from FM waveforms in that the pulse is subdivided into a number of subpulses. The subpulses are of equal duration, and each has a particular phase. The phase of each subpulse is selected in accordance with a given code sequence. The most widely used phase coded waveform employs two phases and is called binary, or biphase, coding. The binary code consists of a sequence of either 0s and 1s or +1s and –1s. The phase of the transmit signal alternates between 0° and 180°. Upon reception, the compressed pulse is obtained by either matched filtering or correlation processing. The width of the compressed pulse at the half amplitude point is nominally equal to the subpulse width. The range resolution is hence proportional to the time duration of one element of the code. The compression ratio is equal to the number of subpulses in the waveform, *i.e.*, the number of elements in the code.

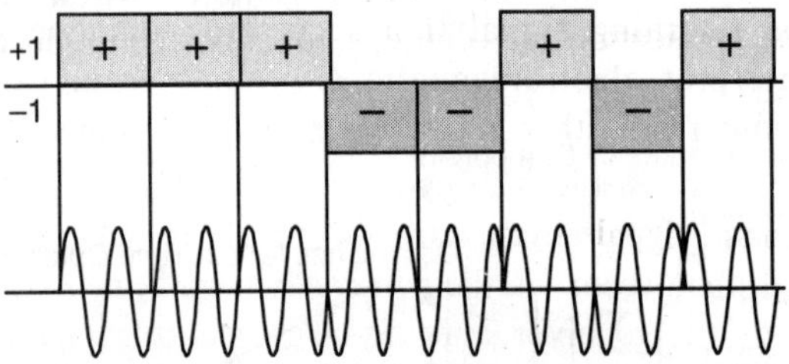

Fig. 6.20. Phase Coded Waveforms.

Factors Affecting Choice of Pulse Compression System

Finally before finishing the discussion of pulse compression, the choice of a pulse compression system is dependent upon the type of waveform selected and the method of generation and processing. The primary factors influencing the selection of a particular waveform are usually the radar requirements of range coverage, Doppler coverage, range and Doppler sidelobe levels, waveform flexibility, interference rejection, and signal to noise ratio (SNR).

6.11. TARGET RECOGNITION

The theory of detection of a target in noise and/or clutter, assuming that the result of a single measurement is a single voltage to be compared with a threshold to determine, whether the target should be declared present of absent. Here, we discuss, an extension of that theory to the case in which several complex voltages are considered simultaneously, in the presence of noise or clutter, and a decision is made regarding whether that set of complex voltages represents a target. More generally, it can represent any set of complex voltages collected by radar, including range versus azimuth for real-beam radar, Doppler bin versus angle, polarimetric or interferometric dimensions and so forth.

Let us, first consider the theoretical problem of deciding whether a set of received complex voltages does or does represent a target of a given type. In reality, even if we know the target type, its representation would depend on its orientation (azimuth and elevation of the radar in target centered coordinates; whether it is articulated (i.e., whether a portion of it is moved relative to its "normal state"); whether it is partially obscured by foliage, nets, weather, or other media; whether it is dirty; and so on. Although actual automatic target recognition (ATR) must contend with those real world.

We have assumed in the probability density distributions for interference (noise and clutter) and signal (target return) were constant. In many cases of interest, the interference background varies with time; in such cases, we usually utilize constant false-alarm rate (CFAR) processing. The threshold level varies with the interference level in such a way as to keep the probability of false alarm constant. In cell averaging CFAR, the threshold for testing for a target at a particular cell is determined by evaluating the mean value of nearby cells.

Radar technology can initially be subdivided into two broad categories. 'Active' Radar transmits an electromagnetic disturbance and then listens for reflections. The time delay between transmission and reception can be used as an accurate estimator of range, if indeed the clock used is sufficiently accurate and stable. Interrogation systems such as IFF are active and also cooperative since target identification information is encoded into the returning signal if a successful dialogue is set up. 'Passive' systems aim to interpret electromagnetic emissions from external sources. For example, transmissions from other active Radars or emissions from a target aircraft.

Radar, like any other sensor, is in essence an information gathering machine. The received electromagnetic signals from active or passive devices will contain a wealth of information imparted upon them via their interaction with the external environment, as well as the effect of the receiver and signal processing chain. To achieve target recognition, a basic minimum amount of information must be distilled from a received signal. For level 1 recognition (as defined in the previous subsection), a target must firstly be detected and discriminated from reflections from thermal noise and unwanted objects (clutter) such as mountains rain and birds. In addition particular features must be extracted from the signal, which are correlated with a distinct target class such that some form of classification can be inferred. It is this latter statement of a requirement of additional information which differentiates a target recognition system from Radar which simply reports detections of wanted objects, and perhaps their location in one or more geometric coordinate.

A summary of such additional information is best aligned to particular technologies, which are tabulated below.

Table 6.2. Summary of Radar based technologies and corresponding measured quantities and target information obtainable

Radar technology	Measured quantities	Possible information obtainable
Active normal radar	Signal to noise ratio Time delay Phase	Range Bearing Elevation Radar Cross-section
Dopper and Interferometic Radar	Phase change of a sequence of returning signals	Range rate Signal modulations from moving parts
Polarimetric Radar	Signal strengths in various polarizations	Indications of the target scattering mechanisms
Plot and Track Extractor	History of target coordinates + uncertainty	Estimate of target position and velocity vectors
Synthetic Aperture Radar (SAR)	As Doppler Radar + accurate (moving) Radar Position	Reflectivity map of surface illuminated by Radar
Inverse SAR (ISAR)	As Doppler Radar (assumes change of target aspect angle)	Target image, with major scatterers highlighted

The hierachy of Radar derived target intelligence: from recognition to identification

There is an important difference between recognition and identification of targets. Although no general standard exists, about this. We describe a four part breakdown of target recognition, with the ultimate goal being full identification.

1. *Recognition of general nature of target*: Broad class distinction such as aircraft, ship, motor vehicle, land clutter, sea clutter etc.
2. *Recognition of target type*: Initial division of broad class. *e.g.,* 'aircraft' can be subdivided into missile, propeller driven, jet, and helicopter, at limits, distinction of targets such as a multi engine commercial airliner from a fighter jet. It is the methodology of such perceptual classification.
3. *Recognition of target class*: This is essentially species identification. For 'aircraft' it would reveal 'Airbus A380', 'Boeing 747-400'.
4. *Full target identification*: Determination of name or other unique identifier of object being interrogated. For example: 'Airbus A340, ABC Airlines flight AM42 from Delhi to Mumbai".

6.11.1. Automatic Target Detection and Recognition Overview

A number of related disciplines (*e.g.,* image enhancement, feature extraction, pattern classification) intersect in support of Automatic Target Detection and Recognition (ATD/R) research. These may be considered and addressed separately, however, when the goal is end-to-end system design for an operational system it is crucial they all be addressed together when predicting final performance metrics. A complete image exploitation system typically consists of five stages plus related enhancement algorithms. Figure 6.21 shows a simple block diagram of radar which is employing ATR system.

The five stages are: Detection, Discrimination, Classification, Recognition, and Identification, as shown in Fig. 6.22. These stages may require a multi-level security approach to properly handle data sensitivities and access.

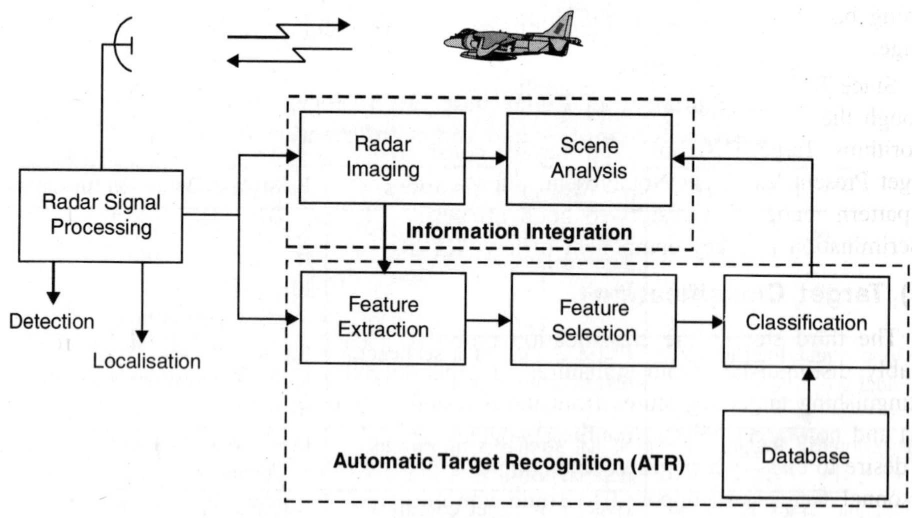

Fig. 6.21. A Simple B/D of Radar with ATR.

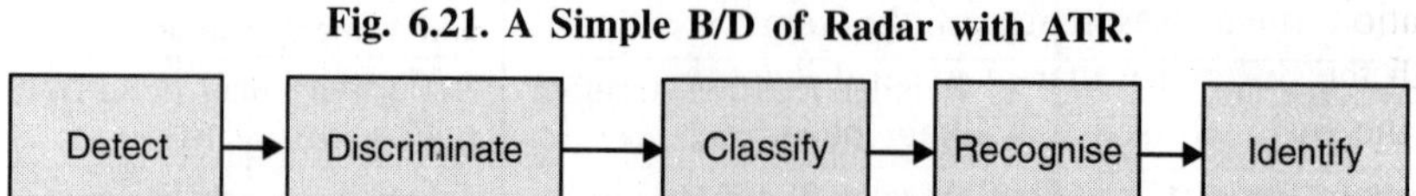

Fig. 6.22. Automatic Target Detection/Recognition Processing Flow.

(i) Detection

The first step is the detection problem, which determines the presence of the target signatures in the sensor data and reliably differentiates targets from clutter. At this stage we are interested in determining and analyzing features of target signatures, which are separable from background. Generally, if the background can be accurately modelled as a stationary random process, the detection problem is comparatively easy. The performance of detection decreases when targets are embedded in non-spatially homogeneous backgrounds or when the sensor data contains other manmade objects. The Probability of Detection (PD) at this first stage must be as close to 100% as possible with the consequence that the Probability of False Alarm Rate is allowed to be comparatively high.

This ensures that targets are not missed. As all image pixels are processed in the detection stage the detection algorithm must be computationally efficient. Typical target detection algorithm consists of a local threshold of pixel intensity, with the threshold set to maintain a specified Probability of Detection (*e.g.*, PD = 90%) and an acceptable Probability of False Alarm or False Alarm Rate (*e.g.*, 10 False Alarms/ Sq Km). Unfortunately, increasing the Probability of Detection comes at the price of increased False Alarm Rate so the first stage detection statistics must be chosen with care.

(ii) Target Discrimination

After the detection stage the False Alarm Rate will be unacceptably high within any real world SAR imagery, thus it will be necessary to perform Target Discrimination in order to reduce the False Alarm Rate prior to Target Classification processing.

False Alarms also arise from manmade returns from windmills, agricultural fences, any metallic object such as a road sign, and commercial vehicular traffic. Ocean surveillance has its own set of phenomena related to the presence of small fishing boats, other unidentified shipping traffic, and sea state effects on the SAR image.

Since Target Discrimination is applied only to Regions of Interest (ROI) that passed through the first Target Detection stage, we may apply more sophisticated processing algorithms. Target Discrimination may be considered binary pattern classification, *i.e.*, Target Present Vs Target Not Present. Such a viewpoint allows one to utilize the field of pattern recognition results to good effect in reducing False Alarm Rate. Target Discrimination is a key component of any ATD/R system.

(iii) Target Classification

The third step is the classification problem. That is, can a target signature be reliably distinguished from signatures of other targets? This problem also includes distinguishing target signatures from those resulting from the clutter (buildings, trees, etc.) and non-target objects (confuser vehicles, etc.). In the Target Classification stage we desire to classify a target as belonging to a general class such as Tank or Armoured Personnel Carrier, as opposed to some non-militarily significant target. The Target Classification stage uses many of the same techniques of the Target Discrimination stage with the difference that in general it is no longer a binary classification problem and that the features used will likely be different.

Although Target Classification is well understood in general and not too difficult for clean imagery, the problem becomes difficult when the targets are camouflaged, located under foliage, revetted, and/or mixed with urban clutter.

(iv) Target recognition

After Target Classification, one also desires to be able to recognise targets by type within a class. Target recognition is defined as automatically recognising a T-72 tank, for example, from the class of targets determined to be military tanks. The ability to do so depends on knowing the various target signatures, there being reasonable differences among those signatures, and having a sensor to be able to sense those differences. The variability of target signatures results from a number of factors. Since target location and orientation are not generally known in advance we must be prepared to detect and recognise targets at all positions and orientations (pose). Alternatively, pose invariant features of target signatures must be developed. Also, targets have parts that can articulate. Large metal pieces can significantly alter targets shapes and signatures. Smaller parts (such as hatches, guns, toolboxes, etc.) also result in variations in target signatures. Determining, measuring and storing all the variations in target signatures is an ongoing task in ATR research and development efforts.

(v) Target Identification

The basic ATR target detection, classification, and recognition stages can be subsequently are followed up by a Target Identification stage. Target Identification refers to the ability to precisely identify a target. For example, after determining that the target is a tank, the tank is a T-72, Target Identification would add that perhaps the T-72 is serial number 1357. It is generally the case that successful Target Identification requires all source data and data fusion techniques.

Fig. 6.23. The MESAR2 S-band experimental multi-function phased array Radar designed and installed at Cowes, Isle of Wight, United Kingdom.

6.12. CLUTTERS

6.12.1. Introduction

Clutter is defined as an unwanted radar echoes. Such echoes may clutter the radar output (*e.g.*, display) and make difficult to detect the desired target signals. When the radar is searching for the targets, clutter echoes received from the hills, trees, high rise structure, land, birds and surface of the sea. Clutter can be due to reflection from storm clouds, clean air tubulance, precipitation and other meteorological phenomena, may confuse the radar operator. However, targets are considered as clutter depending upon the radar applications such as ground echoes might be considered clutter by an airborne interception radar but for the airborne mapping radar, ground return is not clutter. Clutter echoes from the sea, ground or whether are caused by transmission of radar itself. The interference caused by other radar transmission, communication or counter measure transmitter differ in nature from clutter signals. The effect of clutter can be minimised by using MTI radars.

6.12.2. Types of Radar Clutter

Generally radar clutter are being classified as of two types :

(*i*) Ground clutter or surface clutter

(*ii*) Sea clutter.

6.12.2.1. Ground Clutter

Generally, we say the clutters which are appearing on the radar scope due to the reflection from the ground, known as ground clutter. The amount of clutter will depend upon the height of the antenna above the ground. Clutters will be more, as we increase the height of the antenna above the ground. For the ground based radar, clutter signals are mainly from the permanent or fixed targets. Whereas for air borne radar the clutter is varying continuously because of the movement of the aircraft. The echo signals received from the buildings, towers and other man-made structures give more strong echo than other countryside because of the presence of flat reflecting surfaces and corner reflections. River, road and runway backscatter little energy but are visible on radar PPI scopes.

A PPI representation of typical ground echoes might consist of many bright-spots if the beamwidth of the antenna is broad so that individual targets are not resolved. Due to the ever-changing nature of most clutter echoes with time, the conglomeration of spots on the PPI scope displaying clutter may differ from scan-to-scan.

The ground based radars will receive the strong signals from the hills, mountains or other surfaces which are oriented properly. The reflections from hills, and land surfaces are usually very intense that the reflection from the desired target such as aircraft limits the detection capability of the radar. An MTI receiver may be used to minimise the ground clutter and make possible to display the echoes of desired moving targets. The intensity of ground clutter echoes may be further enhanced by super refractions effects.

6.12.2.2. Clutter Cross-sections

While describing the clutter from the ground or sea, the cross-section per unit of intercepted area (σ°) is often taken as a measure of echo strength instead of the more usual cross-section (σ) defined. With an extended target such as clutter, σ is a function of the size of the beam which is illuminating the ground or the sea. Whereas, $\sigma°$ is almost independent of the size of the clutter patch is illuminated. In some cases, the parameter $\gamma = \frac{\sigma°}{\sin\phi}$ has been used to measure the echo-signal return. Where ϕ is the angle of depression of the antenna beam as measured from the horizontal.

If the C is the velocity of propagation and T is the pulse width, all the individual scatterers located within the distance $\frac{CT}{2}$ along the line of propagations results to the composite clutter echo. It can be observed in the figure shown as Fig. 6.24, that $\sigma°$ is equal to $\frac{\sigma}{\left[\left(\frac{CT}{2}\right) R\theta_B \sec\phi\right]}$, where σ is the total cross-section of all the individual scatterers positioned within track of the antenna beam illuminating on the earth's surface, R is the range and θ_B is the half power beamwidth (HPBW) in the azimuth plane.

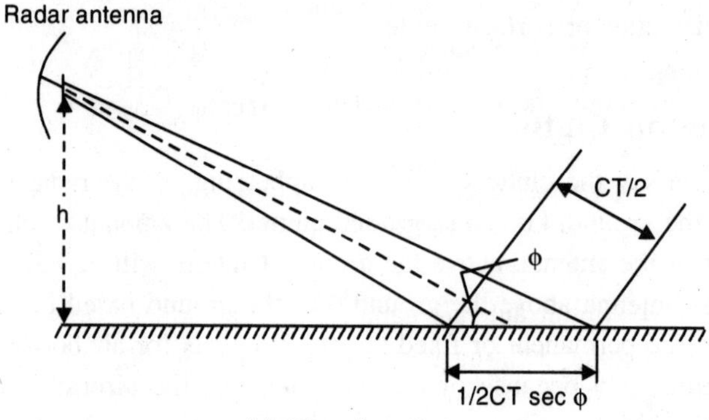

(*a*) Elevation view.

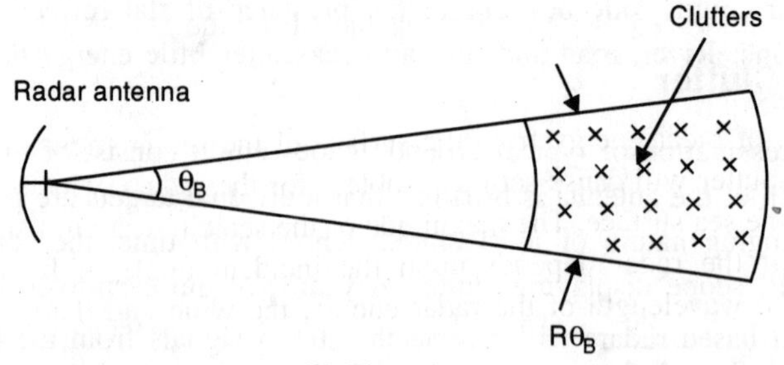

(*b*) Plan view.

Fig. 6.24. Geometrical Representation of Radar Clutters.

It is difficult to describe the specification of the radar cross-section of ground return because of the many different types of terrain and many more factors which are influencing the properties of scattering. The main parameters which are affecting radar scattering from the ground are as follows :

(*i*) Roughness of the surface (*ii*) Incident angle
(*iii*) Polarization (*iv*) Frequency
(*v*) Dielectric constant of the reflecting surface.

It is important to have the knowledge for search radar and air borne interceptions radars in order to predict the amount of echo with which the desired target signal must compete.

6.12.2.3. Antenna Beam

A characteristic of a radar antenna whose radiation pattern obeys a cosecant-squared law is that an isolated, scatter (isotropic) returns an echo signal independent of the range. The cosec^2 antenna provides the constant signal strength with range if following conditions are fulfilled :

(*i*) Isotropic scatter,

(*ii*) Earth is assumed as flat,

(*iii*) There is no reflecting surface which will cause a lobed elevation pattern.

The assumptions that return from ground is from a single scatter and whose radar cross-section is independent of angle is not always true. The radar cross-section per unit of intercepted area, $\sigma°$, is proportional to sin ϕ, where 'ϕ' is the angle of depression. The radar cross-sections σ of rough terrain is therefore proportional to $R \sin \phi / \cos \phi = h / \cos \phi$, not the constant. Where R is the range from radar to the ground target.

$$h = \text{radar height above the ground}$$

$$h = R \sin \phi$$

The antenna gain should be proportional to $\text{cosec}^2 \phi \sqrt{\cos \phi}$

or
$$G(\theta) = G_0 \frac{\text{cosec}^2 \phi \sqrt{\cos \phi}}{\text{cosec}^2 \phi_0 \sqrt{\cos \phi}}$$

for $\phi_0 < \phi < \phi_m$

where G_0 = Maximum antenna gain.

ϕ_0, ϕ_m = Angular defining region over which antenna beam is shaped.

Generally, the shape of the antenna beam designed to produce a uniform ground return is apparently not too critical. Furthermore, that a cosec^2 antenna pattern may be closed to any of the other pattern by varying the tilt of the antenna.

6.12.2.4. Sea Clutter

The reflection of radar energy from the surface of the sea is known as **sea echo** or **sea clutters**. Sea clutter will cause serious problem for the radars, which are positioned or operating near the sea surface. The magnitude of the sea clutter return signal scattered in the direction of the radar depends upon the incident angle with the horizontal, polarisation and the wavelength of the radar energy, the wind and the state of the sea.

Phenomenological Description

We may consider that the surface of the sea having of a number of individual scatters which reflect the EM energy from the radar. The average echo signal reflected from all the independent scatters illuminated by the antenna beam is given by

$$P_r = \frac{P_t G^2 . \lambda^2}{(4\pi)^2 R^4} \sum_i \bar{\sigma}_i \qquad ...(6.24)$$

where $\bar{P}_r$ = Average echo signal

P_t = Transmitted power

G = Gain of the antenna

λ = Wavelength

R = Range

$\bar{\sigma}_i$ = Time averaged radar cross-sections of the ith scatterer

Above eqn. (10.24) is simply derived from the simple form of radar range equations. If σ_0 is defined as the average cross-sections per unit area

$$\sum_i \bar{\sigma}_i = \frac{\sigma^0 R\theta_B CT}{2},$$

then the simple radar equation becomes

$$\bar{P}_r = \frac{P_t G^2 \lambda^2 \theta_B}{(4\pi)^3 R^3} \cdot \frac{CT}{2} \cdot \sigma^0 \quad \text{(for small depression angle)} \quad ...(6.25)$$

where θ_B = Antenna azimuth beamwidth

In the above case the received echo power from the extended clutter may varies inversely as the third power of the range instead of the fourth power relationship obtained for a single point target. This is the result of the extended nature of the clutter. As the range increasing the more the beam will spread and there will be more number of scatterers illuminated.

There is another case of a pencil beam antenna at large depression angles. The pulse packet will cover a much larger region of the sea surface from the area intercepted by the beam. The summation may be determined by the antenna beamwidth in azimuth (θ_B) and in elevations (ϕ_B) rather than the width of the pulse.

Now if ϕ is the angle of depression, we may write

$$\sum_i \bar{\sigma}_i = \frac{\sigma^0 R^2 \theta_B \phi_B}{\sin \phi}$$

and

$$\bar{P}_r = \frac{P_t G^2 \lambda^2 \theta_B \phi_B \sigma^0}{(4\pi)^3 R^2 \sin \phi} \quad \text{(large depressions angle)} \quad ...(6.26)$$

The Eqns. (6.25) and (6.26) are the phenomenological descriptions of sea clutter and they describe the actual phenomenon but avoid explanations.

6.12.3. Clutter Reduction

Clutter reductions techniques are to be used to minimise the effect of clutter in the radar. Normally echo signal from the ground, sea or weather are usually unwanted because they observe the desired targets except in some cases where radar is being used to map terrain. There are some clutter reduction techniques which are as follows :

(*a*) One method of discriminating a moving target in the presence of clutter is to take advantages of Doppler shift in frequency resulting due to motion of the target. This is the case of MTI radar which is already discussed in chapter 3.

(*b*) Another technique is employed which is less sophisticated means for alternating the clutter signal as well as desired signal by simply reducing the gain at close ranges. This is known as Sensitivity Time Control (STC).

(*c*) **Matched filter for clutter reduction :** This techniques of clutter reduction is based on the matched filter principle. Clutter reduction techniques employing the Doppler frequency shift or the matched clutter rejection filter can increase the target signal to clutter-signal ratio.

(*d*) A receiver with a logarithimic or a limiting characteristics to prevent receiver saturation. STC and the log receiver reduce the receiver gain when clutter echoes are present to prevent relatively large clutter signals from saturating the receiver.

6.12.4. Angels

Sometime, radar echoes can be received from the regions of the atmosphere where no apparent reflecting sources exist. These echoes are known as "phantonis", "ghosts", "pixies" and "angels". The most popular name for these type of radar echoes is "Angels". Angels echoes take many different form and have been attributed to various causes, including insects, birds and weather effect. The most important source of angels for ground based radar looking over the sea and shipboard radars operating near the coast, is the "birds". Even though the radar cross-sections of a single bird is small compared with that of an ordinary aircraft, the echoes received from the birds can be relatively strong especially at the shorter ranges became of the inverse form power variation with the range.

Insects, even though very small may be the size of housefly may be readily detected by radar. The radar cross-section of an insects is on the order of 10^{-5} mV at wavelength of 8.6 mm. Usually insects are carried by the wind, therefore angel due to insects might be expected to have the velocity of the wind. Insects and birds echoes are more likely to be found at the lower altitudes, near down and twilight.

Insects can impair the proformances of microwave radar in addition to causing unwanted echoes on radar displays, if they enter the transmission line such as waveguide.

It has been observed that all the angel activity cannot be due to presence of birds and insects, even they may appear to be the result of meteorological effects. Angels do not seam to occur when the ground is covered with snow or when the atmosphere is extremely cold or dry.

Angels echoes need not be a serious source of confusion to the radar operator because usually they are weak and occur at the shorter ranges.

SUMMARY

- Some characteristics of the target that might be extracted from the radar signals. These are as follows :

  (*i*) Range of the target (*ii*) Doppler velocity
  (*iii*) Angle of elevation and azimuth (*iv*) Size of the target
  (*v*) Shape of the target, and (*vi*) Change of shape of the target.
- There are six possible first derivatives from which the nature of the target can be analysed.
- Measurements of target shape, size and rotation may be obtained by measuring the variation of the amplitude with the same dimensions.
- Observation of the phase of the received signals at the antenna as a function of position, when time and frequency remain constant, determine *the angular direction of the target.*
- The variation of phase with frequency, when time and position remain constant then it is a measure of *the range of the target.*
- The variations of signal amplitude with position, when the frequency and time are constant, is related to the *shape of the target.*

- The six measurement derivatives are as follow :

  (*i*) $\left(\frac{\partial \phi}{\partial x}\right)_{t, f} \longrightarrow$ Angle

  (*ii*) $\left(\frac{\partial \phi}{\partial f}\right)_{t, x} \longrightarrow$ Range

  (*iii*) $\left(\frac{\partial \phi}{\partial t}\right)_{x, f} \longrightarrow$ Relative velocity

  (*iv*) $\left(\frac{\partial A}{\partial x}\right)_{tf} \longrightarrow$ Shape

  (*v*) $\left(\frac{\partial A}{\partial t}\right)_{x, f} \longrightarrow$ Change of shape

  (*vi*) $\left(\frac{\partial A}{\partial f}\right)_{t, x} \longrightarrow$ Size

- Three measurements made by radar are as follow :

  (*i*) The range to the target

  (*ii*) The rate of change of range

  (*iii*) The angle of arrival

- A technique of height finding employed is the **'V' Beam Radar.** This consists of two fan shape beams, are vertical and the other is tilted at some angle to the vertical.
- The height of the target can be find by following relation,

$$h = \frac{R \sin \Delta\omega}{(1 + \sin^2 \Delta\omega)^{1/2}}$$

- In short, narrow pulse widths are desirable, but they are not always feasible. Pulse compression radar make use of specific signal processing techniques to provide most of the advantages of extremely narrow pulse widths whilst remaining within the peak power limitations of the transmitter. Pulse compression is the compromise transmits a long pulse that has a bandwidth corresponding to a short pulse.
- Pulse Compression is the processing of a wideband, coded signal pulse, of initially long time duration and low range resolution, to result in an output pulse of time duration corresponding to the reciprocal of the bandwidth and, hence, higher range resolution, and with approximately the same pulse energy.
- There are two basic ways to implement intra pulse modulation, coding either the frequency or the phase of the transmitted pulses, resulting in frequency coded or phase coded waveforms.
- The basic methods of implementation of pulse compression are analog and digital pulse compression techniques.
- The ability of the receiver to improve the range resolution over that of the conventional system is called the ***pulse compression ratio (PCR).***
- The compression ratio is equal to the number of sub pulses in the waveform, *i.e.*, the number of elements in the code.

- Analog Pulse Compression involves the use of analog methods to generate and process pulse compression waveforms.
- The main techniques incorporate active devices, primarily oscillators and passive devices, primarily different delay lines for linear FM waveforms.
- Digital Pulse Compression involves use of digital technology for generating and processing pulse compression waveforms. It has some distinct advantages over analog pulse compression technique.
- *Frequency Modulated [FM] Pulse Compression* is one of the basic techniques for improving the range resolution of radar while maintaining adequate duty factor and average power.
- The five stages of target recognition are: Detection, Discrimination, Classification, Recognition, and Identification.
- **Clutter** is defined as an unwanted radar echoes. Such echoes may clutter the radar output (*e.g.*, display) and make difficult to detect the desired target signals.
- Clutter reductions techniques are to be used to minimize the effect of clutter in the radar.
- Sometime, radar echoes can be received from the regions of the atmosphere where no apparent reflecting sources exist.

REVIEW QUESTIONS

1. What do you understand by measurements of radar parameters ?
2. What are the various characteristics of the target that are to be extracted from the radar signals ?
3. Explain the measurements of phase and amplitude of radar signal.
4. What are the various phase derivatives ?
5. Explain the amplitude derivatives.
6. Explain the measurement of height of a target.
7. What do you understand by range accuracy ?
8. What do you understand by accuracy of frequency ?
9. Explain pulse compression technique. What are the various methods are used to get pulse compression ?
10. Explain FM modulated waveform used in pulse compression.
11. Why pulse compression is needed in the radar?
12. What do you understand by target recognition? Explain the simple block diagram of ATR.
13. What is clutter? How we can reduce the clutter's affects in radar.
14. What are the clutter reduction techniques are used in radar?

❑❑❑

APPENDICES

APPENDIX

A

GLOSSARY

A

Acquisition Mode

That mode of operation wherein the Target Tracking Radar (TTR) scans its narrow beam over an angular segment to initially acquire the target so that the track mode can be used.

Acquisition Radar

Those radars in an air defense network used to locate a target within an angular segment, generally 5-10 degrees.

Airborne MTI (AMTI)

MTI on airborne platform is generally more critical than ground based operation, due to the worse clutter illumination geometry. Tight control on ground sidelobes is generally required.

Airborne Fire Control Radars

Those radars in an air defense network carried on board interceptor aircraft. Their small size and short range characterize them as high frequency, low power, and relatively large beam (3-degree) systems.

Ambiguities

A phenomenon that leads to an erroneous interpretation of target parameters, typically range and doppler. These two parameters are generally conflicting: in pulsed radars, avoiding range ambiguities requires to reduce the PRF, while it should be increased to avoid doppler ambiguity.

Antenna

Device to radiate electromagnetic energy on transmission by radar, and to collect such energy during reception. An antenna pattern is designed with spatial directivity, which concentrates the energy into a beam in both the vertical (elevation) and the horizontal (azimuth) directions. The electrical losses of an antenna together with its directivity determine the antenna gain. In general, the beamwidth in any plane is inversely proportional to the aperture width in that plane, and directly proportional to the wavelength of the radiation. Polarization on transmit and on receive is determined by the antenna.

A-Trace

Amplitude-vs.-range presentation on radar display. Used in tracking radars, usually in conjunction with R-Trace.

Attenuation

Decrease in the strength of a signal, usually described by a multiplicative factor in the mathematical description of signal level. A signal is attenuated by application of a gain less than unity. Common causes of attenuation of an electromagnetic wave include losses through absorption and by volume scattering in a medium as a wave passes through.

AWACS

The AWACS (Airborne Warning and Control System) is a system mounted in a Boeing E3 aircraft that provides airborne surveillance, command and control functions.

Azimuth

The relative position of an object within the field of view of an antenna in the plane intersecting the moving radar's line of flight. The term commonly is used to indicate linear distance or image scale in the along-track direction.

Azimuth Resolution

Resolution characteristic of the azimuth dimension, usually applied to the image domain. Azimuth resolution is fundamentally limited by the doppler bandwidth of the system. Excess doppler bandwidth is usually used to allow extra looks, at the expense of azimuth resolution.

B

Bandwidth

A measure, according to a standard definition, of the span of frequencies available in a signal or other distribution, or of the frequency limiting stages in the system. Bandwidth is a fundamental parameter of any imaging system, and determines the ultimate resolution available. For any pulse, the basic parameter that describes its structure is the time bandwidth product.

Beamwidth

A measure, according to a standard definition, of the width of the radiation pattern of an antenna. Beamwidth may be measured in the one-way or two-way form, and in either voltage or power.

Bistatic

Radar having transmitter and receiver located at different sites.

Blind Range

Range corresponding to an echo delay of one or more PRIs (Pulse Repetition Interval) the echo then arrives at the receiver while the radar is transmitting a new pulse and the receiver is blanked.

Blind Speed

Speed which produce doppler shift, which are integer multiples of the radar PRF, which are therefore aliased to zero doppler and cancelled by the MTI filter.

In other words, speed of a target for which MTI response is zero is called the blind speed.

Brightness

Property of a radar image (digital or optical) in which the observed strength of the radar reflectivity is expressed as being proportional to a digital number (digital image file) or to a gray scale mapping, which, for a photographic positive, shows "bright" as "white".

C

C-Band

Frequency range from 4.0 to 8.0 GHz, according to the "classic" frequency designation.

Chirp

A (generally) linear frequency modulation of the carrier frequency applied within the radar pulse to increase its bandwidth and therefore the range resolution.

Clutter

Unwanted echo (generally from the ground surface), which can obscure targets.

Constant False-Alarm Rate (CFAR)

Condition of a constant probability of false alarms (due to noise and/or clutter) independently of the external conditions.

CFAR condition can be achieved by using adaptive threshold detection logic. Over sea clutter (or gaussian clutter in general) a log-FTC receiver allows to achieve the CFAR condition.

Coherency

Capability of a system to keep a stable phase reference during the target illumination time in order to properly exploit the echo phase information for MTI, synthetic aperture or other purposes.

Coherent-on-Receive

A system using a power oscillator instead of a power amplifier as transmitter (*e.g.*, magnetron), transmitting each pulse with a random phase. Coherency is therefore recovered by locking the receiver phase (usually with a COHO) at each transmitted pulse. Memory of the phase (and therefore coherency) from the previous PRI is therefore lost.

COHerent Oscillator (COHO)

Generally referred to the oscillator working at IF frequency and used as a reference for I/Q demodulation in the Phase Detector. In coherent-on-receive systems, the COHO was used to reconstruct the system coherency by locking its phase to the phase of each downconverted-transmitted pulse.

Conical Scan

A technique used in tracking radars to measure target off-boresight angle in elevation and azimuth. Derived from sequential lobing technique, but here the beam is continuously rotating around the antenna boresight. The target resulting amplitude modulation carries the desired information and, demodulated in its sin and cosine components, provide the angular errors on the two axes.

Cosec^2 Antenna

An antenna with the elevation beam pattern shaped according to the cosec^2 of the elevation angle, often used in 2-D radars. With this pattern shape, a target at a given altitude returns always the same echo, independently of the range.

Command Guidance

A data link whose purpose is the transmission of information from one system to another system.

Continuous Wave Radar (CW Radar)

Continuous wave (CW) radar continually transmits energy in the direction of the target and receives back reflection of the continuous wave.

D

Doppler Beam Sharpening

A technique to enhance the azimuth resolution from a moving platform by discriminating the different doppler shifts from stationary targets at different angles with respect to the direction of motion. Often referred to as "unfocused SAR".

Destaggering

Processing step used in pulse-doppler radars to extract the correct range of targets from range data at different (staggered) PRFs.

Doppler

Shift in frequency of a wave due to the relative motion between the transmitter and the receiver. Frequency shift is 2*v*/lambda. Radars echo are shifted twice this value because this shift must be accounted for both for the forward and for the return path.

Detection

Acquisition of an electromagnetic signal with the same output characteristics as the original transmitted data

Doppler (Effect)

Continuous wave (CW) doppler radar modules are sensors which measure the shift in frequency created when an object (being "illuminated" by energy waves) moves. A transmitter emits energy at a specific frequency which, when reflected, can indicate both speed and direction of that target.

Decibel (dB)

Measurement of signal strength, properly applied to a ratio of powers. For a signal power P compared by ratio to a reference power P_{ref}, the formal definition is $P_{db} = 10 \log_{10} (P/P_{ref})$ Example power ratio of 1/2 corresponds to "–3 dB", derived from $\log_{10} (0.5) = 0.3010$. Decibels often are used in radar, such as in measures of reflectivity, for which the dynamic range may span several factors of ten. The unit is named in honour of Alexander Graham Bell, inventor of the telephone.

Detection

Processing stage at which the strength of the signal is determined for each pixel value. Detection removes phase information from the data file. The preferred detection scheme uses a magnitude squared method, $|s|^2$, which is energy conserving, and has units of voltage squared per pixel.

Dynamic Range

A description of the variety of signal amplitudes (or power levels) available in a system, or present in a data file. Dynamic range is specified either (*i*) to be within minimum and maximum values, or (*ii*) with respect to the ratio of maximum to minimum values. The most important specification is linear dynamic range over which signals combine according to the property of linearity.

E

Electromagnetic (EM) Wave

A wave described by variations in electric and magnetic fields, elegantly formulated by J. C. Maxwell in 1873. Light waves, radio waves, and microwaves are well known examples. All such waves propagate at the speed of light in "free space", which includes most realistic atmospheric conditions. Three material parameters are necessary and sufficient to describe EM waves in a given medium: dielectric constant (or permittivity); permeability; and conductivity.

Energy

For a waveform of time-limited duration such as a radar pulse reflected by an object, the pulse energy is given by the power of the signal integrated over the duration of the signal (Units of watt-seconds = joules).

Electronic Warfare

A general term used to describe the use of communications systems in warfare. As such, EW includes the Electronic Order of Battle (EOB), reconnaissance, intentional interference, intrusion or intelligence collection.

Electronic Counter Measures (ECM)

The intentional use of electronics equipment for the purpose of interference, or confusion in order to obtain a tactical advantage in support of a larger operation. The intention of ECM is to deny or degrade the enemy's use of his communications systems (including radar) in order to gain a time or position advantage. Two basic types exist: passive measures and active measures.

Electronic Counter-Counter Measures (ECCM)

The design or redesign of equipments to make a communication system or equipment techniques less vulnerable to a known or projected ECM equipments.

Electronic Support

Electronic Support aims to gain sufficient information about radar sensors to allow an understanding of the radar's characteristics including its role, its method of operation, and its strengths and weaknesses.

F

False Alarm

False detection due to the noise which, due to its random nature, temporary exceed the detection threshold.

Fan Beam

The "classic" radar antenna beam, with a beam narrow in azimuth and wide enough in elevation to cover all the possible (aerial) targets elevation angles. A radar with this kind of beam cannot discriminate target elevation, *i.e.,* is "bi-dimensional".

Frequency Modulation-Continuous Wave (FM-CW)

A type of radar where a continuous wave instead of pulse is transmitted, in order to avoid doppler ambiguities. The range information is derived by frequency modulating the carrier with a sawtooth waveform and comparing the echo FM modulation envelope with the reference.

Frequency agility

Techniques of rapidly changing the radar carrier frequency (typically, for coherent radars, every processing packet) during the time-on-target, in a random sequence. Commonly used in military radars as an ECCM techniques.

Frequency Modulation (FM)

The modulation of a sine wave carrier so that its instantaneous frequency differs from that carrier by the amount proportional to the modulating wave.

Frequency Spectrum

The entire range of frequencies of electromagnetic radiation.

Fire Control Radar

Specialized radar systems used to locate and track airborne and ground targets to determine optimum weapons firing point and which control the firing and sometimes guidance of weapons

G

Gain

Change in signal level due to processing functions that increase the magnitude of the signal. Examples include: signal amplification in a radar receiver; processing gain in the processor; and antenna gain, a result of the directivity of the pattern.

Gaussian

The classical distribution characterized by a "bell-shaped" curve, it plays several roles in SAR. For example, it is the "normal" probability distribution that describes the in-phase and the quadrature components of the signal corresponding to a purely diffuse scattering surface, which are sometimes described as Gaussian scatterers.

Ground Range

Range direction of a side-looking radar image as projected onto the nominally horizontal reference plane, similar to the spatial display of conventional maps.

For spacecraft data, an Earth geoid model is used, whereas for airborne radar data, a planar approximation is sufficient. Ground range projection requires a geometric transformation from slant range to ground range, leading to relief or elevation displacement, foreshortening, and layover unless terrain elevation information is used.

Ground Penetrating Radar

A type of radar using a low frequency carrier, capable to penetrate the soil and return the echo of buried targets.

H

Height Finder

A type of radar used to determine the elevation of a target detected by a 2-D radar. Uses a nodding-beam antenna pointed on the target azimuth and scanned in elevation.

I

Image

Mapping of the observed radar reflectivity of a scene. For radars with digital image processing, the image consists of a file of digital numbers assigned to spatial positions on a grid of pixels, and presented either as hard copy (such as a photographic print) or soft copy (such as a digital data record). All radar images are subject to statistical variations, mainly speckle and noise, which must be accommodated in either visual or numerical image interpretation. The most commonly used image formats occur after detection. After calibration (and compensation for speckle and noise effects), image files from magnitude squared detection are proportional, on the "standard" for most SAR image files. A magnitude image often yields a photographic copy that is more readily interpreted visually, and requires less dynamic range and data storage space. A digital SAR image file may be retained in complex format (before detection) for specialized applications.

Intensity

Strength of a field or of a distribution, such as an image file, proportional to magnitude, squared.

Interferometer

Device such as an imaging radar that uses two different paths for imaging, and deduces information from the coherent interference between the two signals. In SAR applications, spatial interferometry has been demonstrated to measure terrain height, and time delay interferometry is used to measure movement in the scene such as oceanic currents.

IFF (Identification Friend or Foe)

The identification friend or foe (IFF) system is an example of a secondary radar system that is in wide use in the military environment.

Inverse Synthetic Aperture Radar (ISAR)

Inverse SAR (ISAR) is a variation on the SAR theme that uses the motion of the target relative to the radar to the same effect.

J

Jammer

Device generating some form of intentional interference (noise, swept tone, etc.) to degradate the performance of enemy radars.

K

K-band

Frequency band from 18.0 to 26.5 GHz, according to the "classic" frequency designation

Ka-band

Frequency band from 26.5 to 40.0 GHz, according to the "classic" frequency designation

Ku-band

Frequency band from 12.4 to 18.0 GHz, according to the "classic" frequency designation

L

L-band

Frequency band from 1.0 to 2.0 GHz, according to the "classic" frequency designation. Microwave band in which the wavelengths are at or near 23.5 cm.

Look-Through

During the process of jamming, the operator may interrupt his jamming efforts to determine the enemy's operation mode and to determine if his jamming efforts are effective. If not, he may select a new jamming mode

Leading Edge Tracker

Tracking radar, which obtains its data from the leading edge of the echo pulse from the target.

Laser Radar (LIDAR)

Laser radar (LIDAR) is radar that uses light in place of the radio-frequency signals used in conventional radar.

M

Matched Filter

A receiver filter having a frequency response, which is "matched" of the transmitted pulse (complex conjugated) in order to optimize the signal-to-noise ratio.

For unmodulated pulse, it can be approximated by a bandpass filter of bandwidth=1/T, where T is the pulse width. When a matched filter is used in the receiver, the S/N is proportional only to the pulse energy (P × T) and not to the peak power P.

Missing Detection

Presence of a target, which is not detected by the radar, being the instantaneous value of the signal alongwith noise below the detection threshold.

Multistatic

A radar having one transmitter and several receivers located at different sites.

Monopulse

A technique used in tracking radars to extract the off-boresighth angle (*i.e.*, the X/Y offset w.r.t. the radar electrical axis) four different beams, with a slight pointing offset in X and Y wrt the radar axis, are used. Transmission chain use the sum of these beams, while, on receive; they are combined by sum/subtractions to extract the amplitude difference on the two axis. These differences, normalized in amplitude and phase wrt the sum channel, provide the offset in angle of the target.

Moving Target Indicator (MTI)

An MTI radar is a coherent (both fully coherent or coherent-on-receive) which discriminates moving targets from unwanted fixed target by detecting the pulse-to-pulse phase change wrt an internal phase reference) due to their doppler shift. Non-coherent MTI systems use the clutter phase instead of an internal coherent source.

Multi-Sensor-Tracking

A kind of TWS where the tracking is performed by combining information from more than one radar and/or other sensors.

Maximum Unambiguous Range

The pulse-repetition frequency (PRF) of a radar system determines the maximum operating range of the radar before ambiguities start to occur. This is called the maximum unambiguous range of the radar.

Monostatic Radar

A monostatic radar system transmits and receives its energy through the same antenna system or through collocated antennas.

Missile Guidance Radar (MGR)

The name given to all radars in an air defense network which transmit guidance signals for missiles, whether or not the MGR tracks the missile.

Missile Tracking Radar (MTR)

That radar in an air defense network used to track the missile so that comparison of missile location with TTR data will generate guidance error correction signals which are transmitted by the MTR.

Microwave

An electromagnetic wavelength in (or near) the span 1-100 cm.

N

Noise

Any unwanted or contaminating signal competing with the desired signal.

Noise Figure

Factor that describes the noise level in a radar receiver relative to that in a theoretically perfect receiver. The noise figure, which is always larger than one, is typically two or more, and is usually expressed in decibels.

Nodding-Beam

A nodding beam antenna is basically a fan-beam antenna tilted by 90 degrees, with a beam wide in azimuth and narrow in elevation, and scanned in elevation, used in height finder's radars.

O

OTH, OTH-B

Over The Horizon (Backscatter) A radar working in the HF range (from MHz to tens of MHz) exploiting ionospheric scattering to achieve over-the-horizon performances. Thousands km ranges are possible, with multiple bounces between ionosphere and ground. Performance are strongly dependent on accurate ionosphere behaviour prediction and requires huge antenna complexes.

P

P-Band

The microwave band in which the wavelengths are at or near 75 cm. Frequency band from 0.5 to 1.0 GHz, according to the "classic" frequency designation.

Period

Time duration of one cycle of a wave, or of one cycle of any regularly recurring pattern. Period is inversely equal to frequency. (Units of time, seconds).

Pitch

Vertical rotation of a sensor platform, in the "nose up" plane.

Pixel

Term derived from "picture element" in a digital representation to indicate the spatial position of a sample of an image file, which consist of a spatial array of digital numbers. A two-dimensional ensemble of pixels forms the geometric grid on which an image is built. The fundamental parameter describing this grid is the inter-pixel spacing in each of the two image directions.

Polarization

Orientation of the electric (E) vector in an electromagnetic wave, frequently "horizontal" (H) or "vertical" (V) in conventional imaging radar systems. Polarization is established by the antenna, which may be adjusted to be different on transmit and on receive. Reflectivity of microwaves from an object depends on the relationship between the polarization state and the geometric structure of the object. Common shorthand notation for band and polarization properties of an image file is to state the band, with a subscript for the receive and the transmit state of polarization, in that order. Thus, for example, LHV indicates L-band, horizontal receive polarization, and vertical transmit polarization. Possible states of polarization in addition to vertical and horizontal include all angular orientations of the *E* vector, and time varying orientations leading to elliptical and circular polarizations.

Power

For a given signal, proportional to the magnitude, squared, per unit time. (Units are Watts.)

Propagation

The movement of energy in the form of waves through space or other media. Electromagnetic waves move at the speed c in free space, but the speed v of propagation through other materials is reduced according to the dielectric constant.

Pulse

Group of waves with a distribution confined to a short interval of time. Such a distribution is described in the time domain (or in spatial dimensions) by its width and its amplitude or magnitude, from which its energy may be found. In radar, use is made of modulated or coded pulses, which must be processed, to decode or compress the original pulse to achieve the impulse response observed in the image. Coded pulses have a time-bandwidth product that is much larger than unity. The resolution that may be achieved after processing is determined by the bandwidth of the original pulse.

Passive

Inert components, which may control, but does not create or amplify information for the purpose of jamming.

Phase Coherent

The continuous wave which has no discontinuity.

Phase Locked Loop (P.L.L.)

Circuit in which the local oscillator is synchronized in frequency and phase with the received signal.

Pseudo CW

Method of pulse transmission, which can be received and integrated in a CW receiver as a normal CW signal.

Pulse Coding

A technique that includes a variety of methods to change the transmitted waveform and then decode upon reception.

Pulse Modulation

The modulation of a carrier by a series of pulses generally for the purpose of transmitting data.

Pulse repetition frequency (PRF)

It is the frequency at which a pulse of certain width and amplitude is repeated.

Phased Array Radar

A phased array radar uses an antenna that consists of an array of antenna elements along with signal processing that allows the antenna to be steered electronically.

Plan Position Indicator (PPI)

The plan position indicator (PPI) is possibly the best known of all radar displays where the targets are shown in a plan format. The classic "radar display" providing range-azimuth information on a polar presentation.

Primary Radar

Primary radar systems receive reflections of their own transmitted signals as returned signals from the target.

Pulse Doppler Radar

Pulse doppler radars make use of the Doppler shift in pulse radar to determine the relative velocity of moving targets.

Pulse Radar

Pulse radar transmits a sequence of short pulses of RF energy, and estimates range to the target by measuring the time delay in returned pulses.

Precision Approach Radar (PAR)

A 3-dimensional system composed by two 2-D radars (one with a fan-beam for azimuth and the other with nodding beam for elevation detection) used to guide aircrafts in the final approach to the landing strip.

Probability of False Alarm

Probability that, for each "decision" taken by the radar detector, the noise alone can exceed the threshold and be detected as a target. Higher threshold decrease the PFA at the expense of an increasing probability of missing detection (or a shorter detection range for a given detection probability).

Phase Detector

A device used in radar receiver which provide an output with amplitude proportional to the received signal amplitude and the phase difference between received signal and reference signal. Normally, implemented with mixers, and using two channels, with quadrature reference signal in order to extract sin and cosine component of the received signal.

Pulse Repetition time (prt)

For pulsed Radar, the interval between transmission pulses. It is equal to 1/PRF.

Pulse-Compression

Pulse-compression radars make use of specific signal processing techniques to provide most of the advantages of extremely narrow pulses widths whilst remaining within the peak power limitations of the transmitter. A technique used to achieve a wide pulse bandwidth (and, therefore, range resolution) using long pulse (for high pulse energy with limited peak power) by introducing an intrapulse modulation (*e.g.*, chirp = frequency modulation, Barker = discrete phase modulation) and performing a correlation on the received echo.

Q

Quadrature (Q)

Signal component that is 90° out of phase with respect to the reference frequency.

R

Radar

Electromagnetic sensor characterized by RAdio Detection And Ranging, from which the acronym RADAR is derived. Predicted in the early part of the 20th century, the first important system was built in England in 1938. Basic building blocks of a radar are the transmitter, the antenna (normally used for both transmission and for reception), the receiver, and the data handling equipment. A synthetic aperture radar system, by implication, includes an image processor, even though it may be remotely located in time or space from the radar electronics.

Radar Equation

Mathematical expression that describes the average received signal level (or, sometimes, the image signal level), compared to the additive noise level, in terms of system parameters. Principal parameters include transmitted power, antenna gain, noise power, and radar range R. The range effect is sometimes called the spreading factor, since effective power decreases significantly with a small increase in range.

Radiation

Act of giving off electromagnetic energy, particularly by the antenna of a radar when excited by the transmitter.

Range Ambiguities

Unwanted echoes that fall into the image from ranges that in fact are outside of the intended swath, due to the range sampling operation of the radar. Range ambiguities may be minimized by antenna pattern and imaging mode control, are observed only rarely in imagery from well-designed systems.

Reflectivity

Property of illuminated objects to reradiate a portion of the incident energy. Reflectivity, in general, is larger in the specular direction for smaller surface roughness. For side looking radars, backscatter is the observable portion of the energy reflected. Backscatter, in general, is increased by greater surface roughness. In general, reflectivity is increased for higher conductivity of the scattering surface. The relative strength of radar reflectivity is tabulated by sigma, for discrete objects, and by sigma nought for natural terrain surfaces.

Resolution

Generally (but loosely) defined as the width of the "point spread function", the "Green's function", or the "impulse response function", depending on whether one has an optics, a physics, or an electronic systems background. More properly, "resolution" refers to the ability of a system to differentiate two image features corresponding to two closely spaced small objects in the illuminated scene when the brightness of the two objects in question are comparable and fall within the dynamic range of the radar in question. (Definition adapted from Lord Rayleigh [1879]). "Higher resolution" refers to a system having a smaller impulse response width.

Resolution Cell

A three-dimensional cylindrical volume surrounding each point in the scene. The cell range depth is slant range resolution, its width is azimuth resolution, and its height, which is conformal to the illumination wavefront, is limited only by the vertical beam width of the antenna pattern. Resolution cell often is defined with respect to the local horizontal.

ROLL

Rotation of a sensor platform around the flight vector, hence in a "wing down" direction.

Range Gate Jamming

A deceptive jamming technique used against pulse tracking radars. By varying the delay of the return signal, the range gate is pulled off the true position of the target.

Range Gate Tracker

Radar system, which tracks a target in range by measuring the elapsed time from the transmitted pulse to the echo return.

Range Resolution

The ability of radar to discriminate two targets closely located in range Resolution characteristic of the range dimension, usually applied to the image domain, either in the slant range plane or in the ground range plane. Range resolution is fundamentally determined by the system bandwidth in the range channel.

Range Tracking

Pulse radar measures the time difference between radar pulse transmission and echo reception. The range gate is positioned at a range where the target is expected. The receiver is blanked off except during the period where the range gate is positioned. Range tracking may occur at the leading edge of the return pulse or between ON and OFF gages.

Radar Absorbent Material (RAM)

Radar absorbent material has electrical properties similar to free space. When radar impacts radar absorbent material, the energy acts as though it "sees" infinite free space instead of a boundary.

Radar Altimeter

A radar altimeter is used by an aircraft to determine its height above terrain.

Radar Antenna

A radar antenna is a physical object used to focus, direct and concentrate electromagnetic energy into a specific direction. The most common types of radar antenna are parabolic reflectors and phased array antennas.

Radar Clutter

The term clutter refers to any objects that cause unwanted reflections of a radar's electromagnetic energy to be returned to the radar receiver.

Radar Cross Section (RCS)

The radar cross section (RCS) of a target is the projected area that would intercept the transmitted signal and reflect isotropically an amount that produces the returned signal at the receiver.

Radar Detector

A radar detector warns its user when it detects a radar signal.

Radar Displays

A means of displaying target information in a clear and concise manner.

Radar Range Equation

The radar range equation (RRE) relates the range performance of a radar system to other radar components and their characteristics.

Radar Receivers

Radar receiver is used to amplify and process the reflected signals from the target and make them to display over the screen. Clutter is a major practical problem in the use of radar systems. Radar receivers often incorporate special techniques to minimize clutter.

Radar Reflector

A radar reflector is a passive device that is designed to be an efficient reflector of radar signals, giving a strong return to the radar.

Radar System

A radar system includes: a radar antenna, a radar transmitter, a radar receiver, radar signal processing, and a radar display.

Range

Ambiguous range: range corresponding to a delay greater equal to 1/PRF. If the target is in ambiguous range it is detected as having a smaller range equal to the remainder of its delay divided by (1/PRF), unless ambiguity resolution techniques (such as PRF staggering) are used. An ambiguous echo is also called "Nth time around" echo, where N is the integer result of the aforementioned division (echo of the 1st pulse is detected as if is produced by the Nth pulse).

Slant range : straight range between the radar and the target.

Down Range : range between a (ground) radar and the ground projection of the target (and vice versa for airborne/spaceborn radars)

Range Gate receive "window" opened by tracking radar around the expected range of the target. Range gate position is normally controlled by a closed loop system, which keeps the range gate around the target.

RHI (Range-Height Indicator)

A type of radar display presenting a range/elevation polar plot of the echo of a nodding-beam radar.

S

S-band

Frequency band from 2.0 to 4.0 GHz, according to the "classic" frequency designation.

Synthetic Aperture Radar (SAR)

Radar from a moving platform which performs a coherent summation of the echoes received while illuminating the target from different positions along its path ("SAR baseline") to achieve an angular resolution equivalent to that provided by an antenna having a length equal to the baseline. SAR "focusing" is required to compensate for the phase change due to target range variation during the measurement: unfocused SAR (or "DBS" systems) are limited to smaller baselines with corresponding degraded resolutions.

Scan SAR

A technique to increase the effective swath width of a SAR system beyond the limits imposed by the range/doppler ambiguities trade-off. A smaller swath is electronically scanned in different position to cover the larger swath.

Second-Time-Around Echo

An echo with rank = 2, *i.e.* received after a second pulse is transmitted, and which can be misinterpreted (if no means to solve the ambiguities are available) as an echo from the second pulse at shorter range.

Sequential Lobing

First technique used in tracking radars to extract target off-boresighth information. Four different beams, slightly squinted on.+/-x and +/-y w.r.t. the antenna boresighth, were used sequentially. A target exactly at the centre provided echoes of the same amplitude. An off-boresighth target will provide a stronger return for the beam squinted in the relevant direction. Angular errors extracted in this way were then fed to the tracking servo to keep the target on boresighth.

Single-pass Interferometry

A kind of interferometric SAR operation in which interferometric information are produced with a single pass using two physically spaced antennae.

Slant Range

Absolute range between the radar and the target, measured by means of the echo round-trip time.

SLAR

Side-Looking Airborne Radar. term originally used for airborne reconnaissance radars (initially RARs, then DBS and SAR). RARs needed a long physical antenna which could be only accommodated on aircraft side (therefore looking to the side) while DBS/SARs relies on tangential motion and must therefore "look on sides" by their physical nature.

Spotlight

A technique used in SAR system to improve the angular resolution without increasing the antenna aperture thus reducing its gain, and consisting in the use of a steerable antenna, which is kept, pointed on the target during the synthetic aperture. The improved resolution over stripmap is payed in terms of coverage (only specific "spots" can be mapped in this way, instead of a continuous strip).

Stalo

STAble Local Oscillator. RF oscillator used as 1st local oscillator in the receiver (and for upconversion of Tx signal in fully-coherent systems).

STC

Sensitivity Time Control. Uses an attenuator placed in front of the receiver (usually in the T/R + limiter) to apply an attenuation decreasing with time from the transmit instant, in order to prevent saturation from close-in clutter.

Staggering

To resolve range ambiguities "staggering" of different PRFs is often used. In this way, ambiguous target, illuminated with different PRFs, shows different apparent ranges, allowing to solve the ambiguity and to extract the correct range information.

Secondary Surveillance Radar

A system generally (but not always) associated with search radars, which "interrogates" transponders on board of aircraft and receives their replies (containing information like flight code, altitude and other parameters) for Air Traffic Control purpose. Similar to military IFFs in their mode of operation.

Satellite Radar

Satellite radar systems can be used for a variety of surveillance applications.

Secondary Radar

A secondary radar system is a cooperative target identification system in which the interrogator transmits an encoded signal to a target.

Sidelobe Blanking (SLB)

Antennas with poor sidelobe performance can be improved by using sidelobe blanking techniques.

Side Lobe Control

Side lobe control comes initially from good antenna design where one requirement is the minimization of signals entering the receiver through the sidelobes.

Surveillance Radar

Surveillance radar is used to detect targets within range of the radar.

Scan

The cyclic movement of the beam as a radar examines an angular segment or point.

Stagger PRF

A-radar pulse chain composed of two or more pulse trains of identical PRF.

Signal

Generalized terminology used to signify a mathematical description of a wave, pulse, or other sequence of interest. It often suggests the ensemble of data corresponding to observed scattering from the scene, either before reception, within the radar or processor, or in the image file. Normally there is a distinction between "signal" and noise.

Speed of Light (c)

Approximately 300,000,000 meters per second, the speed of light in "free space", a condition typical of electromagnetic propagation through most atmospheric conditions found on Earth.

T

Track

On search radars, the sequence of discrete "plots" relevant to the same moving targets, scan after scan.

T/R

Transmit/Receive—The device used to perform the duplexing between transmitter and receiver ports when they share the same antenna.

TWS

Track-While-Scan. A technique, which allows to automatically "track" a target from the sequence of plots on a search radar.

TWT

Travelling Wave Tube. Microwave High-power amplifier used as transmitter in fully-coherent radars. Radar system using computer techniques to track targets in range, velocity and position without interferring with the acquisition scan rate.

Transmission

Energy sent by the radar, normally in the form of a sequence of pulses, to illuminate a scene of interest.

Target Tracking Radar (TTR)

That radar in an air defense network used to track the target, thereby obtaining azimuth, elevation and range information from the target and the rates of change of these coordinates so that the radar can keep itself boresighted with the moving target.

Tracking

The continuous monitoring of range, velocity and position of a target in space from a reference position.

W

Wave

Propagating periodic displacement of an energy field. A surface wave on the water serves to visualize the key properties of an electromagnetic wave. At any instant of time, a wave is described by its "height" (amplitude) and its "length" (wavelength). Equally important is the phase of the wave, which is the number that describes the position of the "crests" or "troughs" with respect to a given reference position. At any specific location in space, propagation of the wave occurs. From this perspective, its frequency may be observed. A wave propagates within a given medium at a speed given by the product of its wavelength and its frequency. In radar, waves are very well represented by families of sinusoidal functions, so-called harmonic oscillation.

Wavefront

Three dimensional surface in space for which the field radiated by an antenna has the same phase at all points. At a large distance R from an antenna, the wavefront is a spherical surface with radius R over the angular window established by the antenna pattern. For most geometries encountered in remote sensing, the wavefront may be approximated by a plane tangent to the spherical surface, within a tolerance of much less than a wavelength over a spatial scale of several resolution cells.

Wavelength

Minimum distance between two events of a recurring feature in a periodic sequence, such as the crests in a wave. (Units of length, such as metres).

Weather Radar

Weather radar is used as an aid to weather forecasting.

X

X-Band

Microwave band with wavelength at or near 3 cm. An X band radar operates in the frequency range 8-12 GHz.

Y

YAW

Rotation of a sensor platform in the horizontal plane, hence in a "nose right" direction.

❑❑❑

APPENDIX

B

DECIBEL (dB)

The Decibel is a subunit of a larger unit called the bel. As originally used, the bel represented the power ratio of 10 to 1 between the strength or intensity *i.e.*, power, of two sounds, and was named after Alexander Graham Bell. Thus a power ratio of 10:1 = 1 bel, 100:1 = 2 bels, and 1000:1 = 3 bels. It is readily seen that the concept of bels represents a logarithmic relationship since the logarithm of 100 to the base 10 is 2 (corresponding to 2 bels), the logarithm of 1000 to the base 10 is 3 (corresponding to 3 bels), etc. The exact relationship is given by the formula

$$\text{Bels} = \log\left(\frac{P_2}{P_1}\right) \quad \text{...(1)}$$

where P_2/P_1 represents the power ratio.

Since the bel is a rather large unit, its use may prove inconvenient. Usually a smaller unit, the Decibel or dB, is used. 10 decibels make one bel. A 10:1 power ratio, 1 bel, is 10 dB; a 100:1 ratio, 2 bels, is 20 dB. Thus the formula becomes

$$\text{Decibels (dB)} = 10\log\left(\frac{P_2}{P_1}\right) \quad \text{...(2)}$$

The power ratio need not be greater than unity as shown in the previous examples. In Eqns. (1) and (2), P_1 is usually the reference power. If P_2 is less than P_1, the ratio is less then 1.0 and the resultant bels or decibels are negative. For example, if P_2 is one-tenth P_1, we have

$$\text{Bels} = \log\left(\frac{0.1}{1}\right) = -1.0 \text{ bels}$$

and

$$\text{dB} = 10\log\left(\frac{0.1}{1}\right) = -10 \text{ dB}.$$

It should be clearly understood that the term decibel does not in itself indicate power, but rather is a ratio or comparison between two power values. It is often desirable to express power levels in decibels by using a fixed power as a reference. The most common references in the world of electronics are the milliwatt (mW) and the watt. The abbreviation dBm indicates dB referenced to 1.0 milliwatt. One milliwatt is then zero dBm. Thus P_1 in Eqns. (1) or (2) becomes 1.0 mW. Similarly, The abbreviation dBW indicates dB referenced to 1.0 watt, with P_2 being 1.0 watt, thus one watt in dBW is zero dBW or 30 dBm or 60 dBμW. For antenna gain, the reference is the linearly polarized isotropic radiator, dBLI. Usually the "L" and/or "I" is understood and left out.

dBc is the power of one signal referenced to a carrier signal, *i.e.*, if a second harmonic signal at 10 GHz is 3 dB lower than a fundamental signal at 5 GHz, then the signal at 10 GHz is –3 dBc.

THE DECIBEL, ITS USE IN ELECTRONICS

The logarithmic characteristic of the dB makes it very convenient for expressing electrical power and power ratios. Consider an amplifier with an output of 100 watts when the input is 0.1 watts (100 milliwatts); it has an amplification factor of

$$\frac{P_2}{P_1} = \frac{100}{0.1} = 1000$$

or a gain of:

$$10 \log\left(\frac{P_2}{P_1}\right) = 10 \log\left(\frac{100}{0.1}\right) = 30 \text{ dB}.$$

(notice the 3 in 30 dB corresponds to the number of zeros in the power ratio)

The ability of an antenna to intercept or transmit a signal is expressed in dB referenced to an isotropic antenna rather than as a ratio. Instead of saying an antenna has an effective gain ratio of 7.5, it has a gain of 8.8 dB (10 log 7.5).

A ratio of less than 1.0 is a loss, a negative gain, or attenuation. For instance, if 10 watts of power is fed into a cable but only 8.5 watts are measured at the output, the signal has been decreased by a factor of

$$\frac{8.5}{10} = 0.85$$

or $$10 \log (0.85) = -0.7 \text{ dB}$$

This piece of cable at the frequency of the measurement has a gain of –0.7 dB. This is generally referred to as a loss or attenuation of 0.7 dB, where the terms "loss" and "attenuation" imply the negative sign. An attenuator which reduces its input power by factor of 0.001 has an attenuation of 30 dB. The utility of the dB is very evident when speaking of signal loss due to radiation through the atomsphere. It is much easier to work with a loss of 137 dB rather than the equivalent factor of 2×10^{-14}.

Instead of multiplying gain or loss factors as ratios we can add them as positive or negative dB. Suppose we have a microwave system with a 10 watt transmitter, and a cable with 0.7 dB loss connected to a 13 dB gain transmit antenna. The signal loss through the atmosphere is 137 dB to a receive antenna with a 11 dB gain connected by a cable with 1.4 dB loss to a receiver. How much power is at the receiver? First, we must convert the 10 watts to milliwatts and then to dBm:

$$10 \text{ watts} = 10{,}000 \text{ milliwatts}$$

and $$10 \log\left(\frac{10{,}000}{1}\right) = 40 \text{ dBm}$$

Then 40 dBm – 0.7 dB + 13 dB – 137 dB + 11 dB – 1.4 dB = –75.1 dBm.

–71.1 dBm may be converted back to milliwatts by solving the formula:

$$\text{mW} = 10^{(\text{dBm}/10)}$$

giving: $$10^{(-75.1/10)} = 0.00000003$$

Voltage and current ratios can also be expressed in terms of decibels, provided the resistance remains constant. First we substitute for P in terms of either voltage, V, or current, I. Since $P = VI$ and $V = IR$, we have:

$$P = I^2R = \frac{V^2}{R}$$

Thus for a voltage ratio we have:

$$\text{dB} = 10 \log \left[\frac{\left(\frac{V_2^2}{R}\right)}{\left(\frac{V_1^2}{R}\right)} \right]$$

$$= 10 \log \left(\frac{V_2^2}{V_1^2} \right) = 10 \log \left(\frac{V_2}{V_1} \right)^2$$

$$= 20 \log \left(\frac{V_2}{V_1} \right)$$

Like power, voltage can be expressed relative to fixed units, so one volt is equal to 0 dBV or 120 dBμV.

Similarly for current ratio:

$$\text{dB} = 20 \log \left(\frac{I_2}{I_1} \right)$$

Like power, amperage can be expressed relative to fixed units, so one amp is equal to 0 dBA or 120 dBμA.

Decibel Formulas (where Z is the general form of R, including inductance and capacitance)

When impedances are equal:

$$\text{dB} = 10 \log \frac{P_2}{P_1} = 20 \log \frac{E_2}{E_1} = 20 \log \frac{I_2}{I_1}$$

When impedances are unequal:

$$\text{dB} = 10 \log \frac{P_2}{P_1} = 20 \log \frac{E_2 \sqrt{Z_1}}{E_1 \sqrt{Z_2}} = 20 \log \frac{I_2 \sqrt{Z_2}}{I_1 \sqrt{Z_1}}$$

SOLUTIONS WITHOUT A CALCULATOR

Solution of radar and EW problems requires the determination of logarithms (base 10) to calculate some of the formulae. Common "four function" calculators don't usually have a log capability (or exponential or fourth root functions either). Without a scientific calculator (or math tables or a Log-Log slide rule) it is difficult to calculate any of the radar equations, simplified or "textbook". The following gives some tips to calculate a close approximation without a calculator.

Decibel Table

DB	Power Ratio	Voltage or Current Ratio	DB	Power Ratio	Voltage or Current Ratio
0	1.00	1.00	10	10.0	3.16
0.5	1.12	1.06	15	31.6	5.62
1.0	1.26	1.12	20	100	10
1.5	1.41	1.19	25	316	17.78
2.0	1.58	1.26	30	1,000	31.6
3.0	2.00	1.41	40	10,000	100
4.0	2.51	1.58	50	10^5	316
5.0	3.16	1.78	60	10^6	1,000
6.0	3.98	2.00	70	10^7	3,162
7.0	5.01	2.24	80	10^8	10,000
8.0	6.31	2.51	90	10^9	31,620
9.0	7.94	2.82	100	10^{10}	10

For dB numbers which are a multiple of 10

An easy way to remember how to convert dB values that are a multiple of 10 to the absolute magnitude of the power ratio is to place a number of zeros equal to that multiple value to the right of the value 1.

i.e., 40 dB = 10,000 : 1 (for Power)

Minus dB moves the decimal point that many places to the left of 1.

i.e., –40 dB = 0.0001 : 1 (for Power)

For voltage or current ratios, if the multiple of 10 is even, then divide the multiple by 2, and apply the above rules.

i.e., 40 dB = 100 : 1 (for Voltage)

–40 dB = 0.01 : 1

If the power in question is not a multiple of ten, then some estimation is required. The following tabulation lists some approximations, some of which would be useful to memorize.

You can see that the list has a repeating pattern, so by remembering just three basic values such as one, three, and 10 dB, the others can easily be obtained without a calculator by addition and subtraction of dB values and multiplication of corresponding ratios.

DB Rules of Thumb

Multiply Current/Voltage By			Multiply Power By	
if +dB	if –dB	dB	if +dB	if –dB
1	1	0	1	1
1.12	0.89	1	1.26	0.8
1.26	0.79	2	1.58	0.63
1.4	0.707	3	2	0.5
2.0	0.5	6	4	0.25
2.8	0.35	9	8	0.125
3.16	0.316	10	10	0.1
4.47	0.22	13	20	0.05
10	0.1	20	100	0.01
100	0.01	40	10,000	0.0001

Example 1:

A 7 dB increase in power (3 + 3 + 1) dB is an increase of (2 × 2 × 1.26) = 5 times whereas

A 7 dB decrease in power (–3 – 3 – 1) dB is a decrease of (0.5 × 0.5 × 0.8) = 0.2.

Example 2: Assume you know that the ratio for 10 dB is 10, and that the ratio for 20 dB is 100 (doubling the dB increases the power ratio by a factor of ten), and that we want to find some intermediate value.

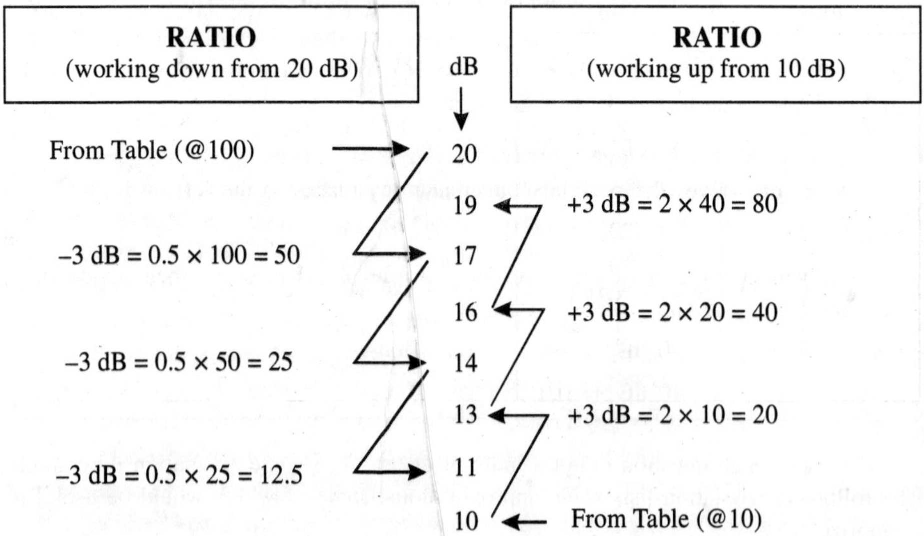

We can get more intermediate dB values by adding or subtracting one to the above, for example, to find the ratio at 12 dB we can:

Work up from the bottom; 12 = 1+11 so we have 1.26 (from table) × 12.5 = 15.75

Alternately, working down the top 12 = 13 – 1 so we have 20 × 0.8 (from table) = 16

The resultant numbers are not an exact match (as they should be) because the numbers in the table are rounded off. We can use the same practice to find any ratio at any other given value of dB (or the reverse).

dB AS ABSOLUTE UNITS

Power in absolute units can be expressed by using 1 Watt (or 1 milliwatt) as the reference power in the denominator of the equation for dB. We then call it dBW or dBm. We can then build a table such as the adjoining one.

From the above, any intermediate value can be found using the same dB rules and memorizing several dB values *i.e.*, for determining the absolute power, given 48 dBm power output, we determine that 48 dBm = 50 dBm – 2 dB so we take the value at 50 dB which is 100 W and divide by the value 1.58 (ratio of 2 dB) to get:

$$\frac{100 \text{ watts}}{1.58} = 63 \text{ W} \quad \text{or} \quad 63{,}291 \text{ mW.}$$

Because dBW is referenced to one watt, the Log of the power in watts times 10 is dBW. The Logarithm of 10 raised by any exponent is simply that exponent. That is: Log $(10)^4 = 4$. Therefore, a power that can be expressed as any exponent of 10 can also be expressed in dBW as that exponent times 10. For example, 100 kW can be written 100,000 watts or 10^5 watts. 100 kW is then +50 dBW. Another way to remember this conversion is that dBW is the number of zeros in the power written in watts times 10. If the transmitter power in question is conveniently a multiple of ten (it often is) the conversion to dBW is easy and accurate.

dB As Absolute Units

dBμW	dBm	Power	dBW
120	90	1 MW	60
90	60	1 kW	30
80	50	100 W	20
70	40	10 W	10
60	30	1 W (1000 mW)	0
50	20	100 mW	–10
40	10	10 mW	–20
33	3	2 mW	–27
32	2	1.58 mW	–28
31	1	1.26 mW	–29

❑❑❑

APPENDIX

C

ABBREVIATIONS AND ACRONYMS

A

a	Acceleration or atto (10^{-18} multiplier)
A	Ampere, Area, Altitude, Angstrom (Å), Antenna Aperture, or Aerial (U.K.)
A-799	No evidence of failure report
A/A, A-A, AA	Air-to-Air or Anti-Aircraft
AA-()	Air-to-Air missile number ()
AAA	Anti-Aircraft Artillery
AAAA	Army Aviation Association of America
AAED	Advanced Airborne Expendable Decoy
AAM	Air-to-Air Missile
AARGM	Advanced Anti-Radiation Guided Missile (concept)
AAW	Anti-Air Warfare
A-BIT	Automatic Built-in-Test
ABM	Air Breathing Missile or Anti-ballistic Missile
A/C	Aircraft (also acft.)
AC	Alternating Current
ACA	Associate Contractor Agreement or Airspace Coordination Area
ACAT	Acquisition Category
ACCB	Aircraft Configuration Control Board
Acft	Aircraft (also A/C)
ACLS	Aircraft Carrier Landing System
ACM	Advanced Cruise Missile or Air Combat Maneuvering
ACQ	Acquisition
ACS	Antenna Coupler Set
ACTD	Advanced Concept Technology Demonstration

A/D	Analog to Digital
Ada	Not an acronym. Ada is the DoD standard programming language.
ADM	Advanced Development Model
ADP	Automatic Data Processing or Advanced Development Program
ADVCAP	Advanced Capability
AEC	Aviation Electronic Combat (Army)
AEGIS	Automatic Electronic Guided Intercept System
AEL	Accessible Emission Limit
AEW	Airborne Early Warning
AF	Antenna Factor, Air Force, or Audio Frequency
AFB	Air Force Base or Airframe Bulletin
AFC	Automatic Frequency Control or Airframe Change
AFIPS	Automated Financial Information Processing System
AFOTEC	Air Force Operational T&E Center
A/G	Air-to-Ground
AGB	Autonomous Guided Bomb
AGC	Automatic Gain Control
AGI	Auxiliary General Intelligence (Intelligence-gathering Ship)
AGL	Above Ground Level
AGM	Air-to-Ground Missile
AGS	Angle Gate Stealer
AHWS	Advanced Helicopter Weapons System
AI	Artificial Intelligence, Air Intercept, or Airborne Interceptor
AIAA	American Institute of Aeronautics and Astronautics
AIC	Air Intercept Control
AIM	Air Intercept Missile
AIRLANT	Commander, U.S. Naval Air Forces, Atlantic Fleet
AIRPAC	Commander, U.S. Naval Air Forces, Pacific Fleet
AJ	Anti-jamming or Anti-Jam
A-Kit	Aircraft wiring kit for a system (includes cabling, racks, etc. excluding WRAs)
AM	Amplitude Modulation
AMD	Aircraft Maintenance Department
AMES	Advanced Multiple Environment Simulator
AMLV	Advanced Memory Loader/Verifier
Amp	Amplifier
AMRAAM	Advanced, Medium-Range, Air-to-Air Missile
ANSI	American National Standards Institute
ANT	Antenna
A_o	Operational Availability

AO	Acousto-Optical
AOA	Angle of Arrival, Angle of Attack, or Analysis of Alternatives (similar to COEA
AOC	Association of Old Crows (Professional EW Society) or Award of Contract
AOT	Angle Only Track, Angle Off Tail, or Acquisition-on-Target
APC	Amphenol Precision Connector or Armored Personnel Carrier
APN	Aircraft Procurement, Navy
APO	Armed Forces (or Army or Air) Post Office, Acquisition Program Office
APU	Auxiliary Power Unit
AR	Anti-reflection or Aspect Ratio
ARM	Anti-radiation Missile
ARO	After Receipt of Order
A/S, A-S, AS	Air-to-Surface
ASCM	Anti-ship Cruise Missile
ASE	Aircraft Survivability (or Survival) Equipment, Allowable Steering Error, or Automatic Support Equipment
ASIC	Application Specific Integrated Circuit
ASK	Amplitude Shift Keying
ASM	Air-to-Surface Missile
ASO	Aviation Supply Office
A-Spec	System Specification
ASPJ	Airborne Self-Protection Jammer
ASPO	Avionics Support (also Systems) Project Office (also Officer)
ASR	Advanced Special Receiver or Airport/Airborne Surveillance Radar
ASRAAM	Advanced Short Range Air-to-Air Missile
ASTE	Advanced Strategic and Tactical Expendables
ASW	Anti-submarine Warfare
ATA	Advanced Tactical Aircraft
ATARS	Advanced Tactical Air Reconnaissance System
ATC	Air Traffic Control
ATD	Advanced Technology Demonstration
ATE	Automatic Test Equipment
ATEDS	Advanced Technology Expendables and Dispenser Systems
ATF	Advanced Tactical Fighter
ATIMS	Airborne Turret Infrared Measurement System or Airborne Tactical Information Management System
ATIRCM	Advanced Threat Infrared Countermeasures

ATP	Acceptance Test Procedure
ATR	Autonomous Target Recognition, Airborne Transportable Rack
ATRJ	Advanced Threat Radar Jammer
AUTODIN	Automatic Digital Network
AUTOVON	Automatic Voice Network (now DSN)
AUX	Auxiliary
avdp.	Avoirdupois (system of measures)
Avg	Average
AWACS	Airborne Warning and Control System

B

B	Bandwidth (also BW) or Magnetic inductance
BAFO	Best and Final Offer
BATBULL	Bat Bulletin - former VX-9 tactics newsletter now called "On Target"
BAU	Bus Adapter Unit
BC	Bus Controller
BDA	Battle Damage Assessment
BDI	Battle Damage Indication
BFO	Beat Frequency Oscillator
BI	Background Investigation
BIFF	Battlefield Identification, Friend, or Foe
BIT	Built-in-Test, Binary Digit or Battlefield Information Technology
BITE	Built-in-Test Equipment
BIU	Bus Interface Unit
B-Kit	Avionics "Black Box" WRAs
B/N	Bombardier/Navigator
BNC	Bayonet Navy Connector
BOA	Basic Ordering Agreement
BOL	Swedish chaff dispenser in a launcher
BPF	Band Pass Filter
BPS	Bits Per Second
BUMED	Bureau of Medicine (Navy)
BUNO	Bureau Number (aircraft)
BUR	Bottom Up Review
BVR	Beyond Visual Range
BW	Beamwidth (referring to an antenna) or sometimes Bandwidth
BWA	Backward Wave Amplifier
BWO	Backward Wave Oscillator

C

c	Speed of Light = 3×10^8 meters/sec = 1.8×10^{12} furlongs per fortnight or 1.8 terafurlongs per fortnight, or centi (10^{-2}) multiplier
C	Electron Charge, Coulomb, Capacitance, Celsius, Centigrade, Confidential, Roman numeral for 100, or a programming language (also C+ and C++)
C^2	Command and Control
C^3	Command, Control, and Communications
C^3CM	C^3-Countermeasures
C^3I	Command, Control, Communications, and Intelligence
CAD	Computer-Aided Design
CAE	Computer-Aided Engineering
CAG	Carrier Air Group
CAGE	Commercial and Government Entry
CAIV	Cost as an Independent Variable
CAL	Calibration
CAM	Computer-Aided Manufacturing or Constant Addressable Memory
CAO	Competency Aligned Organization or Contract Administrative Officer
CAP	Combat Air Patrol
CAS	Close Air Support or Calibrated Airspeed
CASS	Consolidated Automated Support System
CAT	Catapult or Cockpit Automation Technology
CB	Citizens Band (also see Seabee)
CBD	Commerce Business Daily
CBIT	Continuous Built-in-Test
CBO	Congressional Budget Office
CCA	Circuit Card Assembly
CCB	Configuration Control Board
CCD	Charge Coupled Device
CCM	Counter-Countermeasures
CCN	Contract Change Number or Configuration Change Notice
CCU	Cockpit Control Unit
cd	Candela (SI unit of luminous intensity)
CD	Compact Disk or Control and Display
CDC	Combat Direction Center
CDR	Critical Design Review
CDRL	Contract Data Requirements List
CE	Conducted Emission
CECOM	Communications and Electronics Command (Army)

CEP	Circular Error Probability
CFA	Cognizant Field Activity
CFAR	Constant False Alarm Rate
CFE	Contractor Furnished Equipment
CG	Center of Gravity, Commanding General, Command Guidance, or Cruiser
CI	Configuration Item
CIA	Central Intelligence Agency
CIC	Combat Information Center (now called CDC)
CID	Combat Identification or Charge Injection Device
CILOP	Conversion in Lieu of Procurement
CINC	Commander in Chief
CIS	Commonwealth of Independent States (11 of 15 former Soviet Union territories except Estonia, Georgia, Latvia, and Lithuania)
CIWS	Close-In Weapon System
CJ	Coherent Jamming
CLC	Command Launch Computer
cm	Centimeter
CM	Counter Measures or Configuration Management
CMC	Command Mission Computer or Commandant Marine Corps
CMDS	Countermeasure Dispensing System
CMOS	Complementary Metal-Oxide Semiconductor
CMP	Configuration Management Plan
CMWS	Common Missile Warning System
CNAL	Commander, Naval Air Forces Atlantic (COMNAVAIRLANT)
CNAP	Commander, Naval Air Forces Pacific (COMNAVAIRPAC)
CNI	Communications, Navigation, and Identification
CO	Commanding Officer, Contracting Officer, Change Order, or Carbon Monoxide
COB	Close of Business
COEA	Cost and Operational Effectiveness Analysis
COG	Center of Gravity or Cognizant
COMM	Communications
COMSEC	Communications Security
CONSCAN	Conical Scanning Radar
CONUS	Continental United States
CO-OP	Cooperative (countermeasures)
Cos	Cosine
COSRO	Conical-Scan on Receive Only
COTS	Commercial Off-The-Shelf (hardware/software)

CP	Circularly Polarized (antenna), Central Processor, or Command Post
CPS	Computer or Control Power Supply
CPU	Central Processing Unit
CRC	Originally Chemical Rubber Company, now published reference books by CRC Press
CRFM	Coherent RF Memory
CRISD	Computer Resources Integrated Support Document
CRLCMP	Computer Resources Life Cycle Management Plan
CRO	Countermeasures Response Optimization
CRT	Cathode Ray Tube or Combat Rated Thrust (afterburner)
Crypto	Cryptographic
CS	Conducted Susceptibility
CSC	Commodity Software Change
CSCI	Computer Software Configuration Item
C-Spec	Product Specification
CSS	Contractor Support Services
CV	Aircraft Carrier
CVA	Older designation for Attack Carrier
CVN	Nuclear Powered Aircraft Carrier
CVR	Crystal Video Receiver
CW	Continuous Wave or Chemical Warfare
CWBS	Contract Work Breakdown Structure
CWI	Continuous Wave Illuminator
CY	Calendar Year

D

d	Distance, Diameter, or deci (10^{-1} multiplier)
D	Distance, Diameter, Electron displacement, Detectivity, Doppler, Density, or Roman numeral for 500
da	deca (10^{0} multiplier)
D/A	Digital-to-Analog
DAB	Defense Acquisition Board
DAC	Digital to Analog Converter or Dept of Army Civilian
DAR	Defense Acquisition Regulation
DARPA	Defense Advanced Research Projects Agency
DB	Database
dB	Decibel
dBc	dB referenced to the Carrier Signal
dBi	Decibel antenna gain referenced to an isotropic antenna
dBm	Decibel referenced to the power of one milliwatt

DBOF	Defense Business Operations Fund
dBsm	Decibel value of radar cross-section referenced to a square meter
dBW	Decibel referenced to the power of one watt
DC	Direct Current, Discrete Circuit, or District of Columbia
DCE	Data Communication Equipment
DDI	Digital Display Indicator
DDS	Direct Digital Synthesizers
DECM	Deceptive Electronic Countermeasures (also Defensive ECM)
deg	Degree
DEMVAL	Demonstration Validation (also DEM/VAL)
DET	Detachment
DF	Direction Finding
DFT	Discrete Fourier Transform
DI	Data Item
DIA	Defense Intelligence Agency or Diameter
DID	Data Item Description
DIRCM	Directed Infrared Countermeasures
DJ	Deceptive Jamming
D-Level	Depot Level Maintenance
DM	Data Management (also manager)
DMA	Direct Memory Address or Defense Mapping Agency
DME	Distance Measuring Equipment
DNA	Defense Nuclear Agency, Does Not Apply, or Deoxyribonucleic Acid
DOA	Direction of Arrival
DOD or DoD	Department of Defense
DoDISS	DoD Index of Specifications and Standards
DOM	Depth of Modulation
DON	Department of the Navy
DOS	Disk Operating System
DPRO	Defense Plant Representative Office
DRB	Defense Review Board
DRFM	Digital RF Memory
DSARC	Defense Systems Acquisition (and) Review Council
DSN	Defense Switching Network
DSO	Dielectrically Stabilized Oscillator
DSP	Digital Signal Processor
D-Spec	Process Specification
DT (&E)	Development or Developmental Test (and Evaluation)
DTC	Design to Cost

DTE	Data Terminal Equipment
DTO	Digitally Tuned Oscillator or Defense Technology Objectives

E

e	Electron charge or base of natural logarithms (2.71828...)
E	Electric Field Intensity or Strength, Energy, East, or Exa (10^{18} multiplier)
E^3	Electromagnetic Environmental Effects
EA	Electronic Attack (similar to older term of ECM)
EC	Electronic Combat
ECAC	Electromagnetic Compatibility Analysis Center (DOD), now Joint Spectrum Center
ECCM	Electronic Counter-Countermeasures (similar to newer term of EP)
ECL	Emitter Coupled Logic
ECM	Electronic Countermeasures (similar to newer term of EA)
ECN	Engineering Change Notice
ECO	Engineering Change Order
ECP	Engineering Change Proposal or Egress Control Point
ECR	Electronic Combat Range (China Lake) or Electronic Combat & Reconnaissance
ECS	Environmental Control System
ECSEL	Electronic Combat Simulation and Evaluation Laboratory (NAWCWPNS)
ECU	Electronic Control Unit
EDM	Engineering Development Model
EED	Electro-Explosive Device
EEPROM	Electrically Erasable/Programmable Read-only Memory
EHF	Extremely High Frequency (30 to 300 GHz)
EIA	Electronic Industries Associates
EID	Emitter Identification Data
EIRP	Effective Isotropic Radiated Power
EL	Elevation (also El)
ELF	Extremely Low Frequency (3 Hz to 3 kHz)
ELINT	Electronics Intelligence
ELNOT	Emitter Library Notation
EM	Electromagnetic
E-Mail	Electronic Mail
EMC	Electromagnetic Compatibility
EMCAB	EMC Advisory Board
EMCON	Emission Control

EMD	Engineering and Manufacturing Development
EME	Electromagnetic Environment
EMI	Electromagnetic Interference
EMP	Electromagnetic Pulse
EMR	Electromagnetic Radiation
EMS	Electromagnetic Susceptibility
EMV	Electromagnetic Vulnerability
EO	Electro-Optic, Electro-Optical, or Engineering Order
EOB	Electronic Order of Battle or Expense Operating Budget
EOCM	Electro-Optic Countermeasures
EOF	Electro-Optical Frequency (300 to 3×10^7 GHz)
EP	Electronic Protection (similar to older terms of DECM or ECCM)
EPA	Environmental Protection Agency
EPROM	Electrically Programmable Read-only Memory
ERAM	Electronic Counter-Countermeasures (also Protection) Requirements and Assessment Manual
ERP	Effective Radiated Power
ES	Electronic Surveillance (similar to older term of ESM)
ESD	Electrostatic Discharge
ESM	Electronic Support Measures (similar to newer term of ES)
ESSM	Evolved Sea Sparrow Missile
ET	Electronics Technician
ETI	Elapsed Time Indicator
ETR	Estimated Time to Repair
EW	Electronic Warfare or Early Warning
EWAT	Electronic Warfare Advanced Technology
EWIR	Electronic Warfare Integrated Reprogramming (USAF database)
EWMP	Electronic Warfare Master Plan
EWO	Electronic Warfare Officer
EWOPFAC	Electronic Warfare Operational Reprogramming Facility
EWRL	Electronic Warfare Reprogrammable Library (USN)
EWSI	EW Systems Integration
EWSSA	EW Software Support Activity
EXP	Expendable Countermeasure

F

f	femto (10^{-15} multiplier), Frequency (also F), or lens *f* number
F	Frequency (also *f*), Force, Farad, Faraday Constant, Female, Fahrenheit, Noise Figure, Noise Factor or “Friendly” on RWR display
F/A	Fighter/Attack

FAA	Federal Aviation Administration
FAC	Forward Air Controller
FAR	Federal Acquisition Regulations or False Alarm Rate
FAX	Facsimile
f_c	Footcandle (unit of illuminance)
FCA	Functional Configuration Audit
FCR	Fire Control Radar
FDR	Frequency Domain Reflectometry
FEBA	Forward Edge of the Battle Area
FET	Field-Effect Transistor
FEWSG	Fleet Electronic Warfare Support Group
FFT	Fast Fourier Transform
FIFO	First In/First Out
FIPR	Federal Information Processing Resources
fl	fluid
FLAK	AAA Shrapnel, from the German "Flieger Abwher Kanone" (AAA gun that fires fast and furiously)
FLIR	Forward Looking Infrared
FLPS	Flightline Payload Simulator
FLT	Flight
FM	Frequency Modulation or Failure Mode
FME	Foreign Material Exploitation
FMEA	Failure Mode and Effects Analysis
FMS	Foreign Military Sale(s)
FOC	Full Operational Capability
FOD	Foreign Object Damage
FORCECAP	Force Combat Air Patrol
FOT&E	Follow-On Test and Evaluation
FOTD	Fiber Optic Towed Device
FOUO	For Official Use Only
FOV	Field of View
FPA	Focal Plane Array
fps	feet per second
FRACAS	Failure, Reporting, Analysis, and Corrective Actions System
FRB	Failure Review Board
FRD	Functional Requirements Document
FSD	Full Scale Development
FSED	Full Scale Engineering Development
FSK	Frequency Shift Keying
FSU	Former Soviet Union

ft	Feet or Foot
FTC	Fast Time Constant
FTD	Foreign Technology Division (USAF)
FWD	Forward
FY	Fiscal Year

G

g	Gravity (also G)
G	Universal Gravitational Constant (also K), Giga (10 multiplier), Conductance, or Gain
G&A	General and Administrative (expense)
GaAs	Gallium Arsenide
GACIAC	Guidance and Control Information Analysis Center (DoD)
gal	Gallon
GAO	General Accounting Office
GBU	Guided Bomb Unit
GCA	Ground Controlled Approach
GCI	Ground Control Intercept
GENSER	General Service
GEN-X	Generic Expendable
GFE	Government Furnished Equipment
GHz	GigaHertz
GI	Government Issue
GIDEP	Government Industry Data Exchange Program
GIGO	Garbage In/Garbage Out
GOCO	Government Owned Contract Operated
GP	General Purpose
GPI	Ground Plane Interference
GPIB	General Purpose Interface Bus
GPS	Global Positioning System
GSE	Ground Support Equipment

H

h	hours, hecto (10 multiplier), Plank's constant, or height (also H)
H	Height (also h), Henry (Inductance), or Irradiance
HARM	High-speed Anti-Radiation Missile
HAWK	Homing All the Way Killer
HDBK	Handbook
HDF	High Duty Factor

HE	High Explosive
HEF	High Energy Frequency (3×10^7 to 3×10^{14} GHz)
HEL	High Energy Laser
HELO	Helicopter Engineers
HERF	Hazards of Electromagnetic Radiation to Fuel
HERO	Hazards of Electromagnetic Radiation to Ordnance
HERP	Hazards of Electromagnetic Radiation to Personnel
HF	High Frequency (3 - 30 MHz)
HIL	Hardware-in-the-Loop
HOJ	Home-On-Jam
HOL	Higher Order Language
HP-IB	Hewlett-Packard Interface Bus
HP-IL	Hewlett-Packard Interface Loop
HPM	High Powered Microwave
HPRF	High Pulse Repetition Frequency
hr	hour
HSDB	High Speed Data Bus
HUD	Heads-Up Display
HV	High Voltage
H/W	Hardware
HWIL	Hardware-in-the-loop
Hz	Hertz (Cycles per second)

I

i	current (also I)
I	Current (also i), Intensity, Irradiance, Intermediate, or Roman Numeral for One
IADS	Integrated Air Defense System
I&Q	In-Phase and Quadrature
IAS	Indicated Airspeed
IAW	In Accordance With
IBIT	Initiated Built-in-Test
IBU	Interference Blanker Unit
IC	Integrated Circuit
ICD	Interface Control Document
ICMD	Improved Countermeasure Dispenser
ICNIA	Integrated Communication, Navigation, Identification Avionics
ICS	Inverse Conical Scan or Intercommunications System (aircraft)
ICW	In Compliance With

ID	Identification
IDA	Institute For Defense Analysis
IDAP	Integrated Defensive Avionics Program
IDECM	Integrated Defensive Electronic Countermeasures
IEEE	Institute of Electrical and Electronic
IF	Intermediate Frequency
IFF	Identification Friend-or-Foe
IFM	Instantaneous Frequency Measurement
IFR	Instrument Flight Rules
IG	Inspector General
IIR	Imaging Infrared
I-Level	Intermediate Level of Repair (also "I" Level)
ILS	Integrated Logistic Support, Instrument Landing System, or Inertial Locator System
ILSMT	Integrated Logistic Support Management Team
IM	Intermodulation or Item Manager
IMA	Intermediate Maintenance Activity
in	Inch
INEWS	Integrated Electronic Warfare System
INS	Inertial Navigation System
INT	Intensity
I/O	Input/Output
IOC	Initial Operational (also Operating) Capability
IOT&E	Initial Operational Test and Evaluation
IPO	International Projects (Program) Office
IPR	In-Progress/Process Review
IPT	Integrated Product (also Program) Team
IR	Infrared
IR&D	Independent Research and Development
IRCM	Infrared Countermeasures
IRDS	Infrared Detecting System
IREXP	IR Expendables
IRIG-B	Inter-range Instrumentation Group B
IRLS	Infrared Line Scanner
IRS	Interface Requirements Specification, IR Suppression or Internal Revenue Service
IRST	Infrared Search and Track
ISAR	Inverse Synthetic Aperture Radar
ISO	Derived from the Greek "isos" meaning "equal", the official title is International Organization for Standardization

ISP	Integrated Support Plan
ISR	Interference to Signal Ratio (also I/S)
ITU	International Telecommunications Union
IV&V	Independent Validation and Verification
IW	Information Warfare

J

J	Jamming, Radiance, Current Density, or Joules
JAAS	Joint Architecture for Aircraft Survivability
JAFF	Jammer (illuminating) Chaff
JAG	Judge Advocate General
JAMS	Jamming Analysis Measurement System
JASSM	Joint Air-to-Surface Standoff Missile
JAST	Joint Advanced Strike Technology
JATO	Jet Assisted Takeoff or Jammer Technique Optimization
JC2WC	Joint Command and Control Warfare Center
JCS	Joint Chiefs of Staff or Joint Spectrum Center (formerly ECAC)
JDAM	Joint Direct Attack Munition
JED	Journal of Electronic Defense (Published by the Association of Old Crows)
JEM	Jet Engine Modulation
JETS	Joint Emitter Targeting System
JEWC	Joint EW Conference or Joint EW Center (now JC2WC)
JMR	Jammer
JOVIAL	Julius' Own Version of International Algorithmic Language (Air Force computer programming language)
JPATS	Joint Primary Aircraft Training System
J/S	Jamming to Signal Ratio
JSF	Joint Strike Fighter
JSGCC	Joint Services Guidance and Control Committee
JSIR	Joint Spectrum Interference Resolution (signal interference portion of MIJI)
JSOW	Joint Stand-Off Weapon (AGM-154A)
JSTARS	Joint Surveillance Target Attack Radar System
JTCG/AS	Joint Technical Coordinating Group for Aircraft Survivability
JTIDS	Joint Tactical Information Distribution System
JV or J/V	Joint Venture

K

k	kilo (10 multiplier) or Boltzmann Constant

K	Kelvin, Cathode, Universal gravitational constant (also G), or Luminous efficacy
KCAS	Knots Calibrated Airspeed
kg	kilogram
kHz	KiloHertz
KIA	Killed in Action
KIAS	Knots Indicated Air Speed
km	Kilometer
KSLOC	Thousand Source Lines of Code (software)
kt	Knot (nautical miles per hour)
kW	Kilowatt

L

l	length (also L) or liter
L	Length (also l), Loss, inductance, Luminance, or Roman Numeral for fifty
LADAR	Laser Detection and Ranging (i.e., laser radar)
LAN	Local Area Network
LANTIRN	Low Altitude Navigation & Targeting Infrared for Night
LASER	Light Amplification by Stimulated Emission of Radiation
LAT	Latitude (0-90E N or S from equator)
lbs	pounds
LCC	Life Cycle Cost(s)
LCD	Liquid Crystal Display or Lowest Common Denominator
LCP	Left-hand Circular Polarization
LDF	Low Duty Factor
LDS	Laser Detecting Set
LED	Light-Emitting Diode
LEX	Leading Edge Extension
LGB	Laser Guided Bomb
LF	Low Frequency (30 - 300 kHz)
LIC	Low Intensity Combat or Laser Intercept Capability
LISP	List Processing (A programming language used in artificial intelligence)
LLL	Low Light Level (as in LLL TV)
lm	lumen (SI unit of luminous flux)
ln	Natural Logarithm
LO	Local Oscillator or Low Observable
LOA	Letter of Agreement (or Acceptance)
LOB	Line of Bearing (see also AOA)

LOG	Logarithm to the base 10 (also log) or Logistician (afterburner)
LONG	Longitude (0-180E E or W from Greenwich, U.K.)
LOR	Level of Repair
LORA	Level of Repair Analysis
LORAN	Long Range Navigation
LORO	Lobe on Receive Only
LOS	Line-of-Sight
LPAR	Large Phased-Array Radar
LPD	Low Probability of Detection
LPI or LPOI	Low Probability of Intercept
LPRF	Low Pulse Repetition Frequency
LR	Lethal Range
LRA	Line Replaceable Assembly
LRF	Laser Rangefinder
LRIP	Low Rate Initial Production
LRU	Line Replaceable Unit
LSA	Logistic Support Analysis
LSAR	Logistic Support Analysis Record
LSB	Least Significant Bit
LSI	Large Scale Integration
LSO	Landing Signal Officer
LSSO	Laser System Safety Officer
LTBB	Look Through Blanking Bus
LWIR	Long Wave Infrared
LWR	Laser Warning Receiver
lx	Lux (SI unit of illuminance)
LZ	Landing Zone

M

m	milli (10^{-3} multiplier), meter, or electron mass
M	Mega (10^{6} multiplier), Male, Mach number, or Roman numeral for 1,000
MA	Missile Alert or Missile Active
MAD	Magnetic Anomaly Detection (also Detector)
MADD	Microwave Acoustic Delay Device
MAF	Maintenance Action Form
MAG	Marine Aircraft Group or Magnetic
MANPADS	Man-portable Air Defense System
M&S	Modeling and Simulation

MASER	Microwave Amplification by Simulated Emission of Radiation
MATE	Modular Automatic Test Equipment
MAW	Missile Approach Warning System (also MAWS) or Marine Aircraft Wing
MAX	Maximum or Maximum aircraft power
MBFN	Multiple Beam Forming Network
MC	Mission Computer
MCP	Micro-Channel Plate
MDF	Mission Data File
MDI	Multiple Display Indicator or Miss Distance Indicator
MDG	Mission Data Generator
MDS	Minimum Discernible Signal or Minimum Detectable Signal
MDU	Multipurpose Display Unit
MF	Medium Frequency (300 kHz to 3 MHz)
MFD	Multifunction (video) Display
MG	Missile Guidance
MHz	MegaHertz (10 Hz)
MIA	Missing in Action
MIC	Microwave Integrated Circuit or Management Information Center
MICRON	10^{-6} meter
MiG	Mikoyan-Gurevich (Soviet aircraft manufacturer)
MIGCAP	MiG Combat Air Patrol
MIJI	Meaconing, Intrusion, Jamming & Interference (also see JSIR)
mil	One-thousandth of an inch
MIL	Military power (100%, no afterburner) or Military
MILCON	Military Construction
MILSPEC	Military Specification
MILSTRIP	Military Standard Requisitioning and Issue Procedure(s)
MIMIC	Microwave Monolithic Integrated Circuit (also MMIC)
MIN	Minimum
MIPPLE	RWR display switching between ambiguous emitters
MIPS	Millions of (Mega) Instructions Per Second
ML	Missile Launch
MLC	Main Lobe Clutter
MLV	Memory Loader Verifier
MLVS	Memory Loader Verifier Set
mm	Millimeter
MM	Man Month
MMIC	Microwave Monolithic Integrated Circuit (also MIMIC)

MMW	Millimeter Wave (40 GHz or higher per IEEE, but commonly used down to 30 GHz)
MOA	Memorandum of Agreement
MOAT	Missile on Aircraft Test (Phoenix test on F-14)
MOE	Measure of Effectiveness
MOM	Methods of Moments (also MoM) or Metal-Oxide-Metal
MOP	Modulation on Pulse or Measure of Performance
MOPS	Million Operations Per Second
MOS	Minimum Operational Sensitivity, Military Occupational Specialty, Metal-Oxide Semiconductor, or Measure of Suitability
MOSAIC	Modeling System for Advanced Investigation of Countermeasures
MOU	Memorandum of Understanding
MPD	Multi-Purpose Display or Microwave Power Device
MPE	Maximum Permissible Exposure
mph	Miles per Hour
MPLC	Multi-Platform Launch Controller
MPM	Microwave Power Module
MPPS	Million Pulses Per Second
MPRF	Medium Pulse Repetition Frequency
mr or mrad	Milliradian
MRC	Maintenance Requirement Card or Medium Range CAP
MRE's	Meals Ready to Eat
ms	Milliseconds
MSB	Most Significant Bit
MSI	Multi-Sensor (also Source) Integration, Management Support Issues, or Medium Scale Integration
MSIC	Missile and Space Intelligence Center
MSL	Mean Sea Level (altitude) or Missile
MTBF	Mean Time Between Failures
MTI	Moving Target Indicator (or Indication)
MTTR	Mean Time To Repair
MUXBUS	Multiplex Bus
MVS	Minimum Visible Signal
mw	Microwave
mW	Milliwatt
MWIR	Mid Wave Infrared
MWS	Missile Warning Set
MY	Man Year

N

n	nano (10^{-9} multiplier) or number of elements
N	Noise, Newton (force), Radiance, North, or No
n/a	Not Applicable (also N/A)
NA	Numerical Aperture
NAC	Naval Avionics Center (now part of NAWCAD)
NADC	Naval Air Development Center (now part of NAWCAD)
NADEP	Naval Aviation Depot
NASA	National Aeronautics and Space Administration
NATC	Naval Air Test Center (now part of NAWCAD)
NATO	North Atlantic Treaty Organization
NATOPS	Naval Air Training and Operating Procedures Standardization
NAV	Navigation
NAVAIR	Naval Air Systems Command (also NAVAIRSYSCOM)
NAVSEA	Naval Sea Systems Command (also NAVSEASYSCOM)
NAWCAD	Naval Air Warfare Center Aircraft Division (formerly Trenton, NADC, NAC, and NATC)
NAWCWPNS	Naval Air Warfare Center Weapons Division (formerly PMTC, NWC, NWEF, and NOMTS)
NBC	Nuclear, Biological, Chemical
NCTR	Non-Cooperative Target Recognition
NDI	Non-Developmental Item or Non Destructive Inspection
NEI	Noise Equivalent Power
NEMP	Nuclear Electromagnetic Pulse
NEOF	No Evidence of Failure
NEP	Noise Equivalent Power
NF	Noise Figure or Noise Factor (also F)
NFO	Naval Flight Officer
NIPO	Navy International Program Office
NIR	Near Infrared
NISC	Naval Intelligence Support Center
nm	nanometer or Nautical Mile (also NM or NMI)
NM or NMI	Nautical Mile (also nm)
NOHD	Nominal Ocular Hazard Distance
NOMTS	Naval Ordnance Missile Test Station, White Sands, NM (now part of NAWCWPNS)
NORAD	North American Air Defense Command
NPG or NPGS	Naval Post Graduate School
NRE	Non-Recurring Engineering

NRL	Naval Research Laboratory
NRZ	Non Return to Zero
NSA	National Security Agency
nsec or ns	Nanosecond
NSN	National Stock Number
NSWC	Naval Surface Weapons Center
nt	Nit (SI unit of luminance)
NVG	Night Vision Goggles
NWC	Naval Weapons Center (China Lake) now part of NAWCWPNS
NWEF	Naval Weapons Evaluation Facility, Albuquerque, NM (now part of NAWCWPNS)
NWIP	Naval Warfare Information Publication
NWP	Naval Warfare Publication

O

O	Optical
OADR	Originating Agency's Determination Required
OAG	Operational Advisory Group
O&MN	Operations and Maintenance, Navy (also O&M,N)
OBE	Overtaken (Overcome) By Events
OCA	Offensive Counter Air
OEWTPS	Organizational Electronic Warfare Test Program Set
OFP	Operational Flight Program
OJT	On-the-Job Training
O-Level	Organizational Level of Repair (also "O" Level)
OMA	Organizational Maintenance Activity
OMB	Office of Management and Budget
OMEGA	Optimized Method for Estimating Guidance Accuracy (VLF Navigation System)
ONR	Office of Naval Research
OOK	On-Off Keying
OPEVAL	Operational Evaluation
OPM	Office of Personnel Management
OPSEC	Operational Security
OPTEVFOR	Operational Test and Evaluation Force
OR	Operational Requirement or Operationally Ready
ORD	Operational Requirements Document
OSD	Office of the Secretary of Defense
OSHA	Occupational Safety and Health Act

OSIP	Operational Safety Improvement Program
OSM	Operating System Memory or SMA connector made by Omni-Spectra
OT (&E)	Operational Test (and Evaluation)
OTD	Operational Test Director
OTH	Over the Horizon
OTH-B	Over-the-Horizon Backscatter
OTH-R	Over-the-Horizon Radar
OTH-T	Over-the-Horizon Targeting
OTRR	Operational Test Readiness Review
OUSD	Office of the Under Secretary of Defense
oz	ounce

P

p	pico (10^{-12} multiplier) or page
P	Power, Pressure, or Peta (10^{15} multiplier)
P^3I	Pre-Planned Product Improvement
Pa	Pascal (pressure)
PA	Public Address or Program Analyst
PBIT	Periodic Built-in-Test
PC	Pulse Compression, Personal Computer, or Photoconductive
PCA	Physical Configuration Audit
PCM	Pulse Code Modulation
P_d	Probability of Detection
PD	Pulse Doppler
PDI	PD Illuminator or Post Detection Integration
PDP	Plasma Display Panel
PDQ	Pretty Darn (sic) Quick
PDR	Preliminary Design Review
PDW	Pulse Descriptor Word
PEL	Personnel Exposure Limits
PEM	Photoelectromagnetic
PEO	Program Executive Officer
pf	Power Factor or Pico Farads
PFA	Probability of False Alarm
PGM	Precision Guided Munition
ph	Phot (unit of illuminance)
P_h	Probability of Hit
pi	Greek letter π
P_i	Probability of Intercept (also POI)

PID	Positive Identification
PIN	Personal Identification Number
PIP	Product Improvement Plan or Predicted Intercept Point
Pixel	Picture Element
P_k	Probability of Kill or Peak
PLSS	Precision Location Strike System
PM	Phase Modulation or Program Manager
PMA	Program (also Project) Manager, Air
PMAWS	Passive Missile Approach Warning System
PMT	Photomultiplier Tube
PMTC	Pacific Missile Test Center (PACMISTESTCEN) - now part of NAWCWPNS
P-N	Positive to Negative Junction (also p-n)
PN or P/N	Part Number
POC	Point of Contact
POET	Primed Oscillator Expendable Transponder
POI	Probability of Intercept (also PI)
POL	Polarization
POM	Program Objective Memorandum
POP	Pulse-on-Pulse or Product Optimization Program
POST	Passive Optical Seeker Technology (Stinger missile)
PPI	Plan Position Indicator
PPS	Pulses Per Second
PRF	Pulse Repetition Frequency
PRI	Priority or Pulse Repetition Interval
PROM	Programmable Read-only Memory
PRR	Production Readiness Review or Pulse Repetition Rate
PRT	Pulse Repetition Time
P_s	Probability of Survival
P's & Q's	Pints and Quarts (small details)
PSK	Phase-shift Keying
PUPS	Portable Universal Programming System
PV	Photovoltaic
pw or PW	Pulse Width
PWB	Printed Wiring Board

Q

q	electron charge
Q	Quantity Factor (figure of merit), Quadrature, or Charge (coulomb), or aerodynamic pressure

QA	Quality Assurance
QC	Quality Control
QED	Quod Erat Demonstradum (end of proof) (Satirically "quite easily done")
QML	Qualified Manufacturer Listing
QPL	Qualified Parts List
QRC	Quick-Reaction Capability
QRD	Quick Reaction Demonstration
QRT	Quick-Reaction test

R

r or R	Radius or Range or Roentgen
R	Resistance or Reliability
rad	Radian
R&D	Research and Development
RADAR	Radio Detection and Ranging
RADHAZ	Radiation Hazard
RAM	Random Access Memory, Radar Absorbing Material, Rolling Airframe Missile, or Reliability, Availability, and Maintainability
R&M	Reliability and Maintainability
R&R	Rest and Recuperation (Relaxation)
RAT	Ram Air Turbine
RBOC	Rapid Blooming Offboard Chaff
RCP	Right-hand Circular Polarization
RCS	Radar Cross Section
RCVR	Receiver
RDT&E	Research, Development, Test, & Evaluation
RDY	Ready
RE	Radiated Emissions
REC	Receive
RET	Return
RF	Radio Frequency
RFEXP	RF Expendables
RFI	Radio Frequency Interference, Ready-For-Issue, or Request for Information
RFP	Request for Proposal
RFQ	Request for Quotation
RFSS	Radio Frequency Simulation System (Army)
RGPO	Range Gate Pull Off
RGS	Range Gate Stealer

RGWO	Range Gate Walk Off (see RGPO)
RHAW	Radar Homing and Warning Receiver or Radar Homing All the Way
RHAWS	Radar Homing and Warning System
RINT	Radiation Intelligence
RIO	Radar Intercept Officer
RM	Radar Mile
rms or RMS	Root Mean Square
RNG	Range
ROC	Required Operational Capability
ROE	Rules of Engagement
ROI	Return on Investment
ROM	Read-only Memory or Rough Order of Magnitude
ROR	Range Only Radar or Rate of Return (financial)
ROT	Rate of Turn
ROWG	Response Optimization Working Group
RPG	Receiver Processor Group
RPM	Revolutions per Minute
RPT	Repeat
RPV	Remotely Piloted Vehicle
RRT	Rapid Reprogramming Terminal (a type of MLVS)
RS	Radiated Susceptibility or Remote Station
RSDS	Radar Signal Detecting Set
RSO	Range Safety Officer or Receiver, Set-on
RST	Receiver Shadow Time
RT	Remote Terminal, Termination Resistance, or Receiver/Transmitter (also R/T)
RUG	Radar Upgrade
RWR	Radar Warning Receiver
Rx	Receive

S

s, S, or sec	seconds
S	Signal Power, Surface Area, Secret, Electrical conductance (siemens), South, Scattering (as in S-parameters), or Seconds
SA	Situational Awareness, Semi-Active, Spectrum Analyzer, or Surface-to-Air (also S/A or S-A)
SA-()	Surface-to-Air missile number ()
SAE	Society of Automotive Engineers
SAM	Surface-to-Air Missile
SA-N-()	Naval Surface-to-Air missile number ()

SAR	Synthetic Aperture Radar, Special Access Required, Semi-Active Radar, Search and Rescue, or Specific Absorption Rate
SATO	Scheduled Airline Traffic Office
SATS	Semi-Active Test System
SAW	Surface Acoustic Wave
SBIR	Small Business Innovative Research
SCI	Sensitive Compartmented Information
SCIF	Sensitive Compartmented Information Facility
SCN	Specification Change Notice
SCRB	Software Configuration Review Board
SCUBA	Self-Contained Underwater Breathing Apparatus
SCUD	Soviet short-range surface-to-surface missile
SE	Support Equipment
SDLM	Standard Depot Level Maintenance
SDI	Strategic Defense Initiative
Seabee	Someone in the Navy Construction Battalion ("CB")
SEAD	Suppression of Enemy Air Defense (pronounced "seed" or "C add")
SEAL	Sea-Air-Land (Navy special forces)
sec	seconds (also S or s)
SECDEF	Secretary of Defense
SEI	Specific Emitter Identification
SEMA	Special Electronic Mission Aircraft
SERD	Support Equipment Recommendation Data
SHAPE	Supreme Headquarters Allied Powers Europe (NATO military command)
SHF	Super High Frequency (30 to 30 GHz)
SI	Special Intelligence or System International (Units)
SIF	Selective Identification Feature
SIGINT	Signals Intelligence
SIJ	Stand-In Jamming (also S/J)
SIM	Simulation
sin	Sine
SINCGARS	Single Channel Ground and Airborne Radio System
SIRFC	Suite of Integrated RF Countermeasures (includes ATRJ and ATIRCM)
SJ	Support Jamming
S/J	Stand-In Jamming or Signal to Jamming Ratio
SL	Side lobe or Sea Level (also S.L.)
SLAM	Standoff Land Attack Missile
SLAR	Side-Looking Airborne Radar
SLC	Side Lobe Clutter

SLOC	Source Lines of Code or Sea Lines of Communication
SM	Statute Mile (also sm) or Standard Missile
SMA	Scheduled Maintenance Action or Sub-Miniature A connector
SMC	Sub-Miniature C connector
SML	Support Material List
SMS	Stores Management Set or Status Monitoring (sub-) System
S/N or SNR	Signal-to-Noise Ratio
SNORT	Supersonic Naval Ordnance Research Track
SNTK	Special Need to Know
SOF	Safety of Flight
SOJ	Stand-off Jammer
SONAR	Sound Navigation and Ranging atmosphere)
SOO	Statement of Objectives (replacing SOW)
SOP	Standard Operating Procedures
SORO	Scan-on-Receive Only
SOW	Statement of Work (being replaced by SOO)
SPAWAR	Space and Naval Warfare Systems Command
SPEC	Specification
SPIRITS	Spectral Infrared Imaging of Targets and Scenes
SPO	System Program Office
SPY	Radar on an AEGIS ship
sq	Square
sr	Steradian
SRA	Shop Replaceable Assembly
SRAM	Static Random Access Memory
SRB	Software Review Board
SRBOC	Super Rapid Blooming Offboard Chaff
SRD	Systems Requirements Document
SRS	Software Requirements Specification
SRU	Shop Replaceable Unit
SSA	Software (also Special or System) Support Activity, Source Selection Activity, or Solid State Amplifier
SSB	Single Side Band
SSBN	Nuclear Ballistic Missile Submarine
SSGN	Nuclear Guided Missile Submarine
SSI	Small Scale Integration
SSJ	Self Screening Jamming
SSM	Surface-to-Surface Missile
SSN	Nuclear Attack Submarine
SSRO	Sector Scan Receive Only

SSW	Swept Square Wave
S&T	Science and Technology
STANAG	Standardization Agreement (NATO)
STAR	System Threat Assessment Report
stat	Statute
STBY	Standby
STC	Sensitivity Time Control or Short Time Constant or SHAPE Technical Center
STD	Software Test Description, Standard, or Sexually Transmitted Disease
STOVL	Short Takeoff and Vertical Landing
STP	Software Test Plan, or Standard Temperature and Pressure (0°C at 1 atmosphere)
STR	Software (also System) Trouble Report
STT	Single Target Track
STU	Secure Telephone Unit
SUBSAM	Subsurface-to-Air Missile
SUT	System Under Test
S/W	Software (also SW)
SWC	Scan With Compensation
SWM	Swept Wave Modulation
SYSCOM	Systems Command

T

t	Time (also T)
T	Time (also t), tera (10 multiplier), Temperature, or Telsa
TA	Target Acquisition or Terrain Avoidance
TAAF	Teat, Analyze, and Fix
TAC	Tactical Air Command (Air Force)
TACAIR	Tactical Aircraft
TACAMO	Take Charge and Move Out (airborne strategic VLF communications relay system)
TACAN	Tactical Air Navigation
TACDS	Threat Adaptive Countermeasures Dispensing System
TACTS	Tactical Aircrew Combat Training System
TAD	Threat Adaptive Dispensing, Temporary Additional (also Active) Duty, or Tactical Air Direction
T&E	Test & Evaluation
TALD	Tactical Air Launched Decoy
TAMPS	Tactical Automated (formerly Aircraft) Mission Planning System
TAR	Target Acquisition Radar or Training Administrative Reserve

TARPS	Tactical Air Reconnaissance Pod System (used on F-14)
TAS	True Airspeed
TAWC	Tactical Air Warfare Center (Air Force)
TBA	To Be Announced
TBD	To Be Determined
TBMD	Theater Ballistic Missile Defense
TD	Technical Directive (also Director)
TDD	Target Detection Device
TDM	Time Division Multiplexing
TE	Transverse Electric
TEA	Technology Exchange Agreement
TEAMS	Tactical EA-6B Mission Support
TECHEVAL	Technical Evaluation
TEL	Transporter Erector Launcher
TEM	Transverse Electromagnetic
TEMP	Test and Evaluation Master Plan
TEMPEST	Not an acronym. Certification of reduced electromagnetic radiation for security considerations
TERPES	Tactical Electronic Reconnaissance Processing and Evaluation System
TGT	Target
TIM	Technical Interchange Meeting
TM	Telemetry, Transverse Magnetic, or
TMD	Theater Missile Defense
TNC	Threaded Navy Connector
TOA	Time of Arrival
TOJ	Track on Jam
TOO	Target of Opportunity (HARM operating mode)
TOR	Tentative (also Tactical) Operational Requirement or Time of Receipt
TOS	Time on Station
TOT	Time on Target
TOW	Tube-Launched, Optically-Tracked, Wire-guided
TPI	Test Program Instruction
TPS	Test Program Set or Test Pilot School
TPWG	Test Plan Working Group
TQM	Total Quality Management
T/R	Transmit / Receive
TRB	Technical Review Board
TRD	Test Requirements Document
TREE	Transient Radiation Effects on Electronics
TRF	Tuned Radio Frequency

TRR	Test Readiness Review
TS	Top Secret
TSS	Tangential Sensitivity
TSSAM	Tri-Service Standoff Attack Weapon
TT	Target Track
TTI	Time To Impact/Intercept
TTG	Time-to-Go
TTL	Transistor-Transistor Logic
TTR	Target Tracking Radar
TV	Television
TVC	Thrust Vector Control
TWS	Track While Scan or Tail Warning System
TWSRO	Track While Scan on Receive Only
TWT	Travelling Wave Tube
TWTA	Travelling Wave Tube Amplifier
Tx	Transmit
TYCOM	Type Commander

U

u	micron / micro (10 multiplier)
U	Unclassified, Unit, or Unknown (on RWR display)
UAV	Unmanned (also uninhabited) Air (or Aerial) Vehicle
UCAV	Uninhabited Combat Air Vehicle (new Technical Manual USAF term for UAV)
UDF	User Data File
UDFG	User Data File Generator
UDM	User Data Module
UHF	Ultra High Frequency (300 MHz to 3 GHz)
ULF	Ultra Low Frequency (3 to 30 Hz)
μm	Micrometer
UN	United Nations
UNK	Unknown (also U)
UPS	Uninterruptable Power Supply
us or μs	Microseconds
U.S.	United States
USA	United States of America or United States Army
USAF	United States Air Force
USMC	United States Marine Corps
USN	United States Navy
UTA	Uninhabited Tactical Aircraft

UUT	Unit Under Test
UV	Ultraviolet

V

v	Volts (also V), Velocity (also V or v_t)
V	Volts (also v), Velocity (also v or v_t), Volume, or Roman Numeral for five
VA	Veterans Administration, Volt-Amperes, or prefix for a Navy attack squadron
VAQ	Prefix for Navy (or Marine) tactical EW squadron
V&V	Validation and Verification
VCO	Voltage Controlled Oscillator
Vdc or VDC	Volts Direct Current
VDT	Video Display Terminal
VECP	Value Engineering Change Proposal
VF	Prefix for Navy fighter squadron
VFO	Variable Frequency Oscillator
VFR	Visual Flight Rules
VGPO	Velocity Gate Pull Off
VGS	Velocity Gate Stealer
VGWO	Velocity Gate Walk Off
VHF	Very High Frequency (30 - 300 MHz)
VHSIC	Very High Speed Integrated Circuit
VID	Visual Identification
VLF	Very Low Frequency (3 to 30 kHz)
VLSI	Very Large Scale Integration
VLSIC	Very Large Scale Integrated Circuit
VP	Prefix for Navy patrol squadron
VQ	Prefix for Navy special mission (usually reconnaissance) squadron
VRAM	Video Random Access Memory
VS or vs	Velocity Search or Versus (also vs.)
V/STOL	Vertical/Short Take-off and Landing (also VSTOL)
vt	Velocity (also V or v)
VTOL	Vertical Takeoff and Landing
VSWR	Voltage Standing Wave Ratio
VVA	Voltage Variable Attenuator

W

W	Watts, Weight, or West

W&T	Warning & Targeting
WARM	Wartime Reserve Mode
wb	Weber (magnetic flux)
WBS	Work Breakdown Structure
WC	Waveguide, circular
WGIRB	Working Group on Infrared Background
WIA	Wounded in Action
WORM	Write Once Read Many (times) (Refers to optical disks)
WOW	Weight on/off Wheels (also WonW/or WoffW)
WPAFB	Wright-Patterson Air Force Base
WPN	Weapons Procurement, Navy or Weapon
WR	Waveguide, rectangular
WRA	Weapon Replaceable Assembly
WRD	Waveguide, rectangular double ridged
WSSA	Weapons System Support Activity
WVR	Within Visual Range

X

x	Multiplication symbol
X	Reactance, Experimental, Extraordinary, Roman Numeral for ten, or X axis
X-EYE	Cross Eye
XO	Executive officer
X-POL	Cross Polarization
XMIT	Transmit

Y

Y	Yes or Y-Axis
YAG	Yttrium-Aluminum Garnet
yd	Yard
YIG	Yttrium-Iron Garnet

Z

Z	Impedance, Zenith, or Z-Axis
1xLR, 2xLR	One (or two or three etc.) Times Lethal Range
1v1 or 1-v-1	One versus One (Aerial engagement)
2D	Two Dimension
3D	Three Dimension
3M	Navy Maintenance and Material Management System

INDEX

A

B

C

D

E

F

G

H

I

L

M

N

P

R

S

T

W

Z

❑❑❑

Win Prizes

ATTENTION READERS

Win Prizes

We request for your frank assessment regarding various aspects of the book as given below :

INTRODUCTION TO RADAR SYSTEMS

By : K.K. SHARMA

Please Fill up the Sheet in Neat and Clean Writing

(i) What topics of your syllabus, important for your examination point of view, are not covered in this book ?

....

....

....

....

(ii) What are the chapters wherein the treatment of the subject matter is not systematic or organised or updated ?

....

....

....

....

(iii) Have you come across misprints/mistakes/factual inaccuracies in the book ? Please specify the chapters and the page numbers.

....

....

....

....

(iv) What other books on the same subject have you found/heard better than the present book ? Please specify in terms of price and quality.

....

....

....

....

(v) Further suggestions and comments for the improvement of this book.

....

....

....

....

(PS : If need be please attach a separate sheet to write your views)

OTHER DETAILS

(i) Who recommended you this book ? (Please tick (✓) in the box near the option relevant to you.

☐ Teacher ☐ Friends ☐ Bookseller

....

....

....

....

(ii) Name of the recommending teacher, his designation and address:

....

....

....

....

(iii) Name and address of the bookseller you purchased the book from :

....

....

....

....

(iv) Name and address of your institution (Please mention the University or Board as the case may be)

....

....

....

....

(v) Your name and complete address :

....

....

....

....

(vi) Course you are enrolled in :

....

....

....

....

The best assessment will be awarded each month : The award will be in the form of our publications as decided by the Editorial Board.

Please mail the filled up coupon at your earliest to :

S.K. KATARIA & SONS®
4760-61/23, Ansari Road,
Daryaganj, New Delhi-110002
☎ 23243489, 23269324
Fax : 011-23243489
email : katariabooks@yahoo.com